Specific Skills

Division Facts Tips & Tricks

Practice Pages and Classroom Games
for Understanding and Memorizing Facts

by
Barry Doran, Ed.S.
and
Leland Graham, Ph.D.

illustrated by
Vanessa Countryman

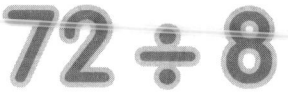

Key Education
An imprint of Carson-Dellosa Publishing LLC
Greensboro, North Carolina

www.keyeducationpublishing.com

CONGRATULATIONS ON YOUR PURCHASE OF A KEY EDUCATION PRODUCT!

The editors at Key Education are former teachers who bring experience, enthusiasm, and quality to each and every product. Thousands of teachers have looked to the staff at Key Education for new and innovative resources to make their work more enjoyable and rewarding. We are committed to developing educational materials that will assist teachers in building a strong and developmentally appropriate curriculum for young children.

PLAN FOR GREAT TEACHING EXPERIENCES WHEN YOU USE EDUCATIONAL MATERIALS FROM KEY EDUCATION PUBLISHING

About the Authors

Barry Doran is a former elementary classroom teacher, mathematics specialist, assistant principal for instruction, Title I coordinator, preK through 12th grade mathematics coordinator, and an adjunct professor at Atlanta Christian College. Mr. Doran retired from the DeKalb County School System in August 2004, having served for 32 years. He served two terms as president of the Georgia Council of Supervisors of Mathematics. In addition to authoring eight educational books, Mr. Doran remains active in mathematics education as a frequent presenter at local, regional, and national conferences.

Dr. Leland Graham is a former college professor, principal, and teacher, who was twice voted "Outstanding Teacher of the Year." The author of 55 educational books, Dr. Graham is a popular speaker and workshop presenter throughout Georgia and the United States, as well as a presenter for NSSEA (National School Supply and Equipment Association). Thousands of teachers have benefited from his workshops on reading, math, and improving achievement scores.

Acknowledgments

The authors would like to thank all of the courageous elementary teachers who teach mathematics every day. Mathematics is the language of the universe—teach our children to speak it fluently.

Credits

Authors: Barry Doran, Ed.S. and Leland Graham, Ph.D.
Publisher: Sherrill B. Flora
Illustrator: Vanessa Countryman
Editors: Debra Olson Pressnall and Karen Seberg
Cover Production: Annette Hollister-Papp
Page Layout: Key Education Staff
Cover Photographs: © Comstock and © ShutterStock

Key Education
An imprint of Carson-Dellosa Publishing LLC
PO Box 35665
Greensboro, NC 27425 USA
www.keyeducationpublishing.com

Copyright Notice

© 2010, Carson-Dellosa Publishing LLC. The purchase of this material entitles the buyer to reproduce worksheets and activities for classroom use only—not for commercial resale. Reproduction of these materials for an entire school or district is prohibited. No part of this book may be reproduced (except as noted above), stored in a retrieval system, or transmitted in any form or by any means (mechanically, electronically, recording, etc.) without the prior written consent of Carson-Dellosa Publishing LLC. Key Education is an imprint of Carson-Dellosa Publishing LLC.

Printed in the USA • All rights reserved.

ISBN 978-1-602680-69-2
01-170128091

Table of Contents

Introduction 4

Teacher and Parent Support

Pretest/Posttest A 5
Pretest/Posttest B 6
Letter to Parents: Learning
    the Division Facts 7
Strategies for Introducing the
    Division Facts 8
Multiplication/Division Table 9
Strategies for Memorizing the
    Division Facts 10–12
Division Facts Review 13
More Division Facts Review 14
Using Manipulatives to Learn
    the Division Facts 15
100's Chart 16

Getting Started

Division by Subtracting 17
Subtract or Divide? 18
Making Equal Groups 19
Circling Equal Groups 20
Think and Draw Equal Groups 21
More Think and Draw Equal Groups 22
Skip Counting Backwards 23
Skip to the Quotient 24
Picture the Fact Families 25
Smart Thinking About Fact Families 26

Division Models

Matching Fact Equals Matching Fact 27
Check the Division Facts 28
A Trick for Even-Number Divisors 29

Division Facts

Dividing by 0 and 1 30
Dividing by 2 and 5 31
More Dividing by 2 and 5 32
Divisible by Doubles 33
Crossword Quotients 34

Picture These Quotients (÷3 and ÷4) 35
Swimming with Facts (÷3 and ÷4) 36
Shade In and Solve (÷6 and ÷7) 37
More Shade In and Solve (÷8 and ÷9) 38
Hit the Mark! (÷6, ÷7, ÷8, ÷9) 39
Read and Solve 40
Smart and Easy Dividing by 10 41

Division Review

Tic-Tac-Divide 1, 2, 5 42
Leapin' Lily Pad Math (÷1, ÷2, ÷5) 43
Climb to the Top (÷10) 44
Ohio Star Quilt Block (÷1, ÷2, ÷5, ÷10) 45
On Your Mark, Get Set, Divide!
    (÷3 and ÷4) 46
Solve the Case of the Missing Numbers
    (÷3 and ÷4) 47
Lift Off with the Facts! (÷6 and ÷9) 48
Catch Air with Facts (÷7 and ÷8) 49
Rev Up for Division (÷6, ÷7, ÷8, and ÷9) 50
More Read and Solve 51
Dive into 45 (Review 45 facts) 52
Burst Past 45 (Review more facts) 53

Classroom Division Games

Directions for Games:
    The Nickel Trick Game 54
    Division in a Bag 54
    Baseball Division Bingo 54
    Division Bowling 55
    Circle Division Game 55
    Division War! 55
    Put a Spin on It! 56
    Number Cube Division 56
Division in a Bag Game Cards 57
Baseball Division Bingo Board 58
Put a Spin on It! Game Materials 59
Tic-Tac-Division 60

Other Resources

Answer Key 61–63
Web Sites 63
Correlations to NCTM Standards 64

Introduction

The acquisition of computational skills is a key component of the mathematics curriculum. The National Council of Teachers of Mathematics (NCTM) in its *Principles and Standards for School Mathematics* (2000) advocates computational fluency, meaning that students are able to compute efficiently and accurately. In September 2006, NCTM released *Curriculum Focal Points for Prekindergarten through Grade 8 Mathematics* in which specific topics of study are outlined for each grade level. One of the grade 3 focal points is developing understanding of division and strategies for basic division facts.

In reading, the goal is for students to be able to read for meaning and recall words quickly and with fluency. The same goal applies to the acquisition of division facts in mathematics. Students should be able to recall facts from memory with quickness and accuracy, and they should also know what those facts mean. In order to achieve that goal, students must have mastery of counting skills, including skip counting backwards by 2s, 3s, 4s, 5s, and 10s. Oral counting can be taught through the use of rhythmic counting, songs, raps, and CDs. A lack of counting skills makes the initial work with division difficult.

The activities and games in *Specific Skills: Division Facts Tips & Tricks* are aligned to both the NCTM Standards and Focal Points. The activities can be used for individual practice, whole group instruction, homework, or enrichment. The activities cover the following key concepts:

- Skip counting backwards
- Repeated subtraction
- Making equal groups
- Connecting multiplication to division
- Specific strategies for memorizing the division facts
- Division practice

The main goal of *Specific Skills: Division Facts Tips & Tricks* is for all students, including reluctant learners, to memorize and master the division facts through 10. The authors advocate using the 11's and 12's facts to introduce two-digit by two-digit division. This strategy will set the stage for working with the division of larger numbers.

A special section for teachers and parents on pages 10–14 outlines a suggested sequence for introducing the facts in a logical order that builds on students' prior knowledge of skip counting backwards and repeated subtraction. The student activities and games provide meaningful practice to aid in the mastery of the basic division facts. Breaking down the facts into smaller "chunks" increases the probability of success for all students. At the end of the book is a list of helpful Web sites for teachers, parents, and students. The sites can be used for lesson planning by teachers and include fun games for students to play.

Name _____   Date _____

Pretest/Posttest A

Directions: Read each problem. Circle the correct answer.

---

1. Which number completes the pattern?

    18, 15, 12, ___

    A. 10   **B.** 9   **C.** 11

2. How many times can you subtract 5 from 15?

    Show your work.

    A. 5   **B.** 4   **C.** 3

3. Read the clues and choose the correct fact.

    12 pencils in total

    3 equal groups

    4 pencils in each group

    A. 3 × 4 = 12   **B.** 12 ÷ 4 = 3
    C. 12 ÷ 3 = 4

4. What is the quotient?

    18 ÷ 2 = ___

    A. 9   **B.** 18   **C.** 36

5. What does the picture show?

    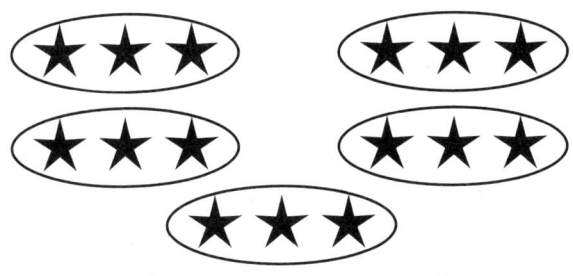

    A. 3 ÷ 5 = 15
    B. 15 ÷ 5 = 3
    C. 5 ÷ 3 = 15

6. Find the quotient.

    7)‾49

    A. 6   **B.** 9   **C.** 7

KE-804077 © Key Education — 5 — Specific Skills: Division Facts Tips & Tricks

Name _____ Date _____

Pretest/Posttest B

Directions: Read each problem. Circle the correct answer.

1. In this pattern, you are counting backwards by what number?

    56, 48, 40, 32

    A. 6   **B.** 8   **C.** 7

2. In the family of facts below, which division fact is missing?

    $3 \times 4 = 12$
    $4 \times 3 = 12$
    $12 \div 3 = 4$
    _____

    A. $12 \div 3 = 4$ **B.** $4 \div 3 = 12$
    C. $12 \div 4 = 3$

3. What number comes next?

    20, 15, 10, ___

    A. 0   **B.** 5   **C.** 2

4. What is the quotient?

    $36 \div 9 =$ ___

    A. 6   **B.** 5   **C.** 4

5. What does the picture show?

    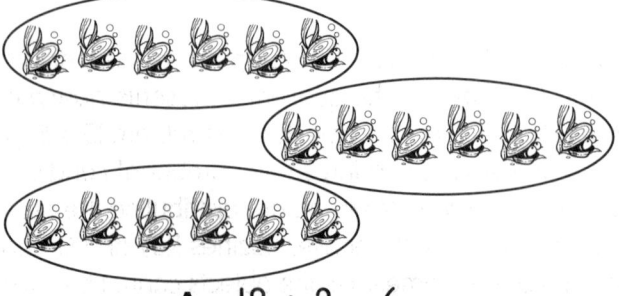

    A. $18 \div 3 = 6$
    B. $3 \div 6 = 18$
    C. $6 \div 3 = 18$

6. What is the quotient?

    $6 \overline{)48}$

    A. 9   **B.** 8   **C.** 6

KE-804077 © Key Education — 6 — Specific Skills: Division Facts Tips & Tricks

> The following parent letter will provide you with the instructional methodologies of how division is currently being taught in our schools.

Learning the Division Facts

Dear Parent or Guardian,

Over the course of several weeks, our class will be learning about division. Students will participate in fun hands-on activities and completing activity sheets to learn the division facts. To help your child better understand these math concepts, please consider practicing the following activities and reading the books detailed later in this letter.

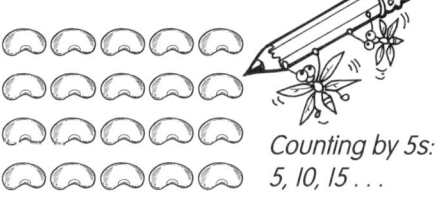
Counting by 5s: 5, 10, 15 . . .

A first step toward understanding division is to demonstrate its relationship with multiplication. You can do this very easily by building models with common objects found in your kitchen, such as dried beans, toothpicks, paper cups, crackers, candy pieces, and pasta shapes. Begin the hands-on session with your child by choosing a multiplication fact, such as 4 x 5. Then, make an array. This is done by arranging the chosen objects (beans) in four rows of five. With your child, skip count by multiples of 5 as you point to each group saying, "5, 10, 15, 20" and call out the answer again (20). Continue the lesson; pile up the beans and introduce the related division fact: 20 ÷ 5. Make up a simple story for your chosen division fact and have your child *remove groups of five beans from the pile to find out how many small equal groups can be made* (4). Repeat this "game" with other multiplication and division facts. It is recommended that you start with the 2's and 5's facts and then proceed with the 3's and 4's facts. While your child removes a specified number of items each time (actually using subtraction) to find the answer for the division problem, point out that the *divisor* stands for how many objects are in each group. The *quotient* represents the number of equal groups. After completing a few hands-on division activities with your child, read together *Divide and Ride* by Stuart J. Murphy (HarperCollins, 1997)—a story about the adventure of 11 friends at an amusement park. The characters in the book quickly learn how to apply division skills.

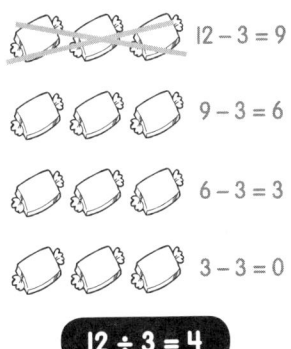

Once your child understands the purpose of subtracting to find the quotient, continue using the strategy of skip counting backwards to figure out the answers for other problems. When your child can easily count by multiples of 2 up to 20 and by multiples of 5 up to 50, then work on counting back to 0. Build a solid understanding of 2's and 5's division facts by having your child memorize them. Your child may depend on the skip counting strategy for awhile. Continue the exercise by reciting the multiples of 3 (3, 6, 9, . . . , 30) and 4 (4, 8, 12, . . . , 40). For example, for the problem 12 ÷ 3, create an array and count backwards by multiples of 3. You would say, "9, 6, 3, 0." (The quotient would be 4.) It will take a lot more practice to recite these multiples and learn the related division facts.

Finally, challenge your child by also creating story problems in which the division process has a different meaning: this time the divisor stands for *the number of equal groups* and the quotient represents *how many objects are in each group.* (The process is known as partition division.) For example, for the problem 15 ÷ 5, you might say, "There are 15 candy pieces and you would like to give each friend the same number of candies. If you have five friends, how many pieces of candy does each one receive?" Now you are distributing the pieces of candy into a specified number of groups. The book *Division with Toys* by Jennifer Rozines Roy and Gregory Roy (Marshall Cavendish, 2007) also shows this process with pictures. Sometimes the objects cannot be evenly divided into groups; then, there is a remainder. The book also addresses this concept.

Thank you for your time and assistance.

Sincerely,

Strategies for Introducing the Division Facts

The most effective instructional strategy for introducing the division facts is to explain the relationship between multiplication and division. This can be achieved by having students investigate families of facts with manipulatives. (If your students are still learning the multiplication facts, you might consider having them complete activities from the book *Multiplication Facts Tips & Tricks* [Key Education Publishing, 2008] to help them recall those facts from memory with quickness and accuracy.)

Once students are able to demonstrate how division is the inverse operation of multiplication, it is time to introduce strategies such as using skip counting backwards, repeated subtraction, and making equal groups to solve simple division problems. The number of division facts that must be memorized can be reduced if students use the strategy of changing division problems into missing-factor multiplication problems and then recall the multiplication facts.

Using the Multiplication Table to Divide

Make a copy of page 9 for each student and/or a transparency of the page for use on an overhead projector. Direct students to look at the multiplication/division table. Review various fact families. For example, introduce the family of facts for 6, 8, and 48. Have students write the two multiplication facts (6 x 8 = 48 and 8 x 6 = 48) on a sheet of paper. Then, let them figure out the two related division facts. (48 ÷ 6 = 8 and 48 ÷ 8 = 6) Show students how division problems can be written as multiplication problems with missing factors. Follow the steps below.

x/÷	0	1	2	3	4	5	6	7	8	9	10
0	-	-	-	-	-	-	-	-	-	-	-
1	0	1	2	3	4	5	6	7	8	9	10
2	0	2	4	6	8	10	12	14	16	18	20
3	0	3	6	9	12	15	18	21	24	27	30
4	0	4	8	12	16	20	24	28	32	36	40
5	0	5	10	15	20	25	30	35	40	45	50
6	0	6	12	18	24	30	36	42	48	54	60
7	0	7	14	21	28	35	42	49	56	63	70
8	0	8	16	24	32	40	(48)	56	64	72	80
9	0	9	18	27	36	45	54	63	72	81	90
10	0	10	20	30	40	50	60	70	80	90	100

Use the Table to Solve a Basic Division Fact
1. Change a division problem into a multiplication sentence. Example: Write 48 ÷ 6 as 6 x ___ = 48.
2. Locate the given product (48) in the row for the known factor (6).
3. Slide your finger up from the given product (48) to find the missing factor (8).
   (Therefore, 6 x 8 = 48 and 48 ÷ 6 = 8.)

Repeat the Steps for the Related Division Fact
To find the other division fact in the given family of 6, 8, and 48, find the 48 circled in the row for the factor 8. The missing factor at the top of that column is 6. So, 8 x 6 = 48 and 48 ÷ 8 = 6.

Name _____   Date _____

Directions on page 8

Multiplication/Division Table

×/÷	0	1	2	3	4	5	6	7	8	9	10
0	*	0	0	0	0	0	0	0	0	0	0
1	*	1	2	3	4	5	6	7	8	9	10
2	*	2	4	6	8	10	12	14	16	18	20
3	*	3	6	9	12	15	18	21	24	27	30
4	*	4	8	12	16	20	24	28	32	36	40
5	*	5	10	15	20	25	30	35	40	45	50
6	*	6	12	18	24	30	36	42	48	54	60
7	*	7	14	21	28	35	42	49	56	63	70
8	*	8	16	24	32	40	48	56	64	72	80
9	*	9	18	27	36	45	54	63	72	81	90
10	*	10	20	30	40	50	60	70	80	90	100

* To focus on division facts, the box was left blank because to divide by 0 is meaningless.

KE-804077 © Key Education — Specific Skills: Division Facts Tips & Tricks

Strategies for Memorizing the Division Facts

The multiplication facts and the division facts should be taught together. The following sequence of strategies can be used to teach the division facts and reinforce the multiplication facts. *Make a copy of pages 13 and 14 for each student and a transparency of each page for use on an overhead projector.* As students learn new facts, have them make their own sets of flash cards. See pages 11 and 12 for additional suggestions.

Dividing by 0 and 1

By definition, division by 0 is meaningless or undefinable.
- Numbers cannot be divided by 0 because it is impossible to make 0 equal groups.

However, 0 divided by any number will always equal 0.
- For example, to check $0 \div 3 = 0$ by using multiplication, the statement $0 \times 3 = 0$ is true.

Any number divided by 1 is that number.
- Whenever a number is divided by 1, there will always be one group and everything is in that group.
- For example, $6 \div 1 = 6$. This means that only one group of 6 objects can be formed.

Have students fill in the charts for the 0's and 1's facts.

Divide by 0
0÷0
1÷0
2÷0
3÷0
4÷0
5÷0
6÷0
7÷0
8÷0
9÷0
10÷0

Divide by 1
0 ÷ 1 = 0
1 ÷ 1 = 1
2 ÷ 1 = 2
3 ÷ 1 = 3
4 ÷ 1 = 4
5 ÷ 1 = 5
6 ÷ 1 = 6
7 ÷ 1 = 7
8 ÷ 1 = 8
9 ÷ 1 = 9
10 ÷ 1 = 10

Dividing by 2, 5, and 10

Before being able to divide by 2, 5, and 10, it is critical that students are able to skip count by multiples of 2, 5, and 10 and have mastered the related multiplication facts. Students can practice skip counting backwards from 20 by multiples of 2, from 50 by multiples of 5, and from 100 by multiples of 10. Keep students engaged by letting them count backwards both orally and on paper. *Have students fill in the corresponding charts on pages 13 and 14 and memorize those facts.*

Counting Backwards by . . .

- **Multiples of 2 (from 20):** 18, 16, 14, 12, 10, 8, 6, 4, 2, 0

- **Multiples of 5 (from 50):** 45, 40, 35, 30, 25, 20, 15, 10, 5, 0

- **Multiples of 10 (from 100):** 90, 80, 70, 60, 50, 40, 30, 20, 10, 0

Divide by 2
0 ÷ 2 = 0
2 ÷ 2 = 1
4 ÷ 2 = 2
6 ÷ 2 = 3
8 ÷ 2 = 4
10 ÷ 2 = 5
12 ÷ 2 = 6
14 ÷ 2 = 7
16 ÷ 2 = 8
18 ÷ 2 = 9
20 ÷ 2 = 10

Divide by 5
0 ÷ 5 = 0
5 ÷ 5 = 1
10 ÷ 5 = 2
15 ÷ 5 = 3
20 ÷ 5 = 4
25 ÷ 5 = 5
30 ÷ 5 = 6
35 ÷ 5 = 7
40 ÷ 5 = 8
45 ÷ 5 = 9
50 ÷ 5 = 10

Divide by 10
0 ÷ 10 = 0
10 ÷ 10 = 1
20 ÷ 10 = 2
30 ÷ 10 = 3
40 ÷ 10 = 4
50 ÷ 10 = 5
60 ÷ 10 = 6
70 ÷ 10 = 7
80 ÷ 10 = 8
90 ÷ 10 = 9
100 ÷ 10 = 10

Divisible by Doubles

There are some division facts that have identical divisors and quotients, such as 16 ÷ 4 = 4 and 36 ÷ 6 = 6. These special facts can be known as the "Doubles Facts." Have students memorize these facts. Point out the facts on the Multiplication/Division Table (see page 9). *There is no separate chart on page 13 or 14 for these facts.*

Divisible by Doubles
1 ÷ 1 = 1
4 ÷ 2 = 2
9 ÷ 3 = 3
16 ÷ 4 = 4
25 ÷ 5 = 5
36 ÷ 6 = 6
49 ÷ 7 = 7
64 ÷ 8 = 8
81 ÷ 9 = 9
100 ÷ 10 = 10

Dividing by 3 and 4

Skip counting backwards can also be used to teach division by 3 and 4. Keep in mind that some students may find this strategy more difficult than skip counting backwards by multiples of 2, 5, and 10. To aid students, have them use copies of the 100's chart on page 16. To count backwards by 3s from 30, have students use their index fingers to point to the next number in the sequence: 27, 24, 21, 18, 15, and so on. Lightly shade each number. Repeat the exercise to count by 4s, counting backwards from 40, and circle the corresponding numbers.

Counting Backwards by . . .

- **Multiples of 3 (from 30):** 27, 24, 21, 18, 15, 12, 9, 6, 3, 0
- **Multiples of 4 (from 40):** 36, 32, 28, 24, 20, 16, 12, 8, 4, 0

Please note: Another effective tool to use when practicing skip counting backwards is a number line.

Have students fill in the corresponding charts on page 13 and memorize those facts.

Divide by 3
0 ÷ 3 = 0
3 ÷ 3 = 1
6 ÷ 3 = 2
9 ÷ 3 = 3
12 ÷ 3 = 4
15 ÷ 3 = 5
18 ÷ 3 = 6
21 ÷ 3 = 7
24 ÷ 3 = 8
27 ÷ 3 = 9
30 ÷ 3 = 10

Divide by 4
0 ÷ 4 = 0
4 ÷ 4 = 1
8 ÷ 4 = 2
12 ÷ 4 = 3
16 ÷ 4 = 4
20 ÷ 4 = 5
24 ÷ 4 = 6
28 ÷ 4 = 7
32 ÷ 4 = 8
36 ÷ 4 = 9
40 ÷ 4 = 10

KE-804077 © Key Education — Specific Skills: Division Facts Tips & Tricks

Dividing by 6, 7, 8, and 9

The strategy of finding the missing factor in a basic multiplication problem can help students memorize the 6's, 7's, 8's, and 9's division facts. Generate various multiplication problems for students to solve and have them use their copies of the Multiplication/Division Table (see page 9) to identify the corresponding division facts. For example, state the problem 8 x __ = 72. Ask students to find the missing factor (9) and tell you the corresponding division fact (72 ÷ 8 = 9).

Next, have students practice by writing two division facts for each missing factor multiplication problem. For example, announce the problem 6 x __ = 42. Students should identify 7 as the missing factor and write 42 ÷ 6 = 7 and 42 ÷ 7 = 6. Repeat the lesson with other problems to introduce all new facts with divisors 6–9.

Have students fill in the corresponding charts on page 14 and memorize those facts.

Divide by 6
0 ÷ 6 = 0
6 ÷ 6 = 1
12 ÷ 6 = 2
18 ÷ 6 = 3
24 ÷ 6 = 4
30 ÷ 6 = 5
36 ÷ 6 = 6
42 ÷ 6 = 7
48 ÷ 6 = 8
54 ÷ 6 = 9
60 ÷ 6 = 10

Divide by 7
0 ÷ 7 = 0
7 ÷ 7 = 1
14 ÷ 7 = 2
21 ÷ 7 = 3
28 ÷ 7 = 4
35 ÷ 7 = 5
42 ÷ 7 = 6
49 ÷ 7 = 7
56 ÷ 7 = 8
63 ÷ 7 = 9
70 ÷ 7 = 10

Divide by 8
0 ÷ 8 = 0
8 ÷ 8 = 1
16 ÷ 8 = 2
24 ÷ 8 = 3
32 ÷ 8 = 4
40 ÷ 8 = 5
48 ÷ 8 = 6
56 ÷ 8 = 7
64 ÷ 8 = 8
72 ÷ 8 = 9
80 ÷ 8 = 10

Divide by 9
0 ÷ 9 = 0
9 ÷ 9 = 1
18 ÷ 9 = 2
27 ÷ 9 = 3
36 ÷ 9 = 4
45 ÷ 9 = 5
54 ÷ 9 = 6
63 ÷ 9 = 7
72 ÷ 9 = 8
81 ÷ 9 = 9
90 ÷ 9 = 10

Flash Card Review

It is imperative that students learn the basic addition, subtraction, multiplication, and division facts. Flash cards offer an excellent way to practice them. For some students this task will be easy to complete, but for others it can be very difficult. Most students will memorize the facts in a short amount of time while others may need lots of skill practice during more than one school year. Flash-card practice sessions can help students initially master the facts as well as provide maintenance through the elementary years.

To help students master the 6's, 7's, 8's, and 9's division facts, have them make their own flash cards using blank index cards. Instead of using the traditional one fact per card format, they should prepare each card by printing a multiplication fact on the front and the two related division facts on the back. This type of flash card practice will help students not only memorize isolated division facts but also see the relationship between multiplication and division.

Let students work in small groups to play a modified game of "War" using the flash cards. Have Player A hold up a multiplication fact card for the group to see. The first student to state the two related division facts correctly wins the card. Player A then holds up another card for the group and so on until all cards have been used. If the other players call out the wrong answer, Player A shows them a new card. The student with the most cards wins the game and becomes Player A for the next game. (If you prefer, the students can write down the two division facts instead of shouting them out. This makes it a much quieter game!)

KE-804077 © Key Education — Specific Skills: Division Facts Tips & Tricks

Name _____ Date _____

Directions on pages 10 and 11

Division Facts Review

Divide by 0	Divide by 1	Divide by 2
~~0 ÷ 0~~	0 ÷ 1 = 0	0 ÷ 2 = 0
~~1 ÷ 0~~	1 ÷ 1 = 1	2 ÷ 2 = 1
~~2 ÷ 0~~	2 ÷ 1 = 2	4 ÷ 2 = 2

Divide by 3	Divide by 4	Divide by 5

Name _____ Date _____

Directions on pages 10–12

More Division Facts Review

Divide by 6
0 ÷ 6 = 0
6 ÷ 6 = 1

Divide by 7

Divide by 8

Divide by 9

Divide by 10

KE-804077 © Key Education — Specific Skills: Division Facts Tips & Tricks

Using Manipulatives to Learn the Division Facts

Skip Counting Backwards Using a 100's Chart

Before focusing on the division facts, spend time with students investigating multiples, including skip counting backwards to 0 (zero). A 100's chart is an excellent tool for teaching this concept.

Give each student a laminated copy of the chart on page 16 and a watercolor marker. First, direct students to shade in box 100 and skip count backwards by 2s, circling the multiples. When everyone is finished, ask students to count the circles that were drawn. (50) Then, continue the lesson to find out how many groups of 5 can be made. Have students start again at 100, skip count backwards by 5s, and draw a square around each multiple. Count the squares that were drawn. (20) Explain that 100 can be divided into 50 equal groups of 2 and also into 20 equal groups of 5.

Have students wipe off their charts and then repeat this exercise as desired to practice identifying the multiples of 3, 4, 6, 7, 8, or 9. For example, direct students to shade in box 30 and skip count backwards by 3s, circling the multiples. Explain that each skip backwards by 3 from 30 represents the subtraction of 3 from a total. In other words, 30 – 3 = 27, 27 – 3 = 24, 24 – 3 = 21, and so on. This process would be repeated 10 times to show 30 ÷ 3 = 10.

Repeated Subtraction Using Pencil, Paper, and Pasta

Pasta shapes, like macaroni or whatever type of pasta you have on hand, are inexpensive and easy to use when demonstrating division concepts. Instead of counting by multiples, have students use repeated subtraction to find the number of equal groups that can be formed. For example, give students 28 pieces of pasta. Ask them to subtract 7 and keep subtracting a group of 7 until no pieces remain. Count the groups of pasta that have been made (4). Show the math sentences on a piece of paper, such as:

28 – 7 = 21   21 – 7 = 14   14 – 7 = 7   7 – 7 = 0
Division sentence: 28 ÷ 7 = 4

Making Equal Groups Using Pasta

Another way to interpret division is to find out how many objects will be in a group (known as partition division). Provide pasta shapes and let students solve problems. Each time, have students draw a given number of circles on their papers and then solve to find out how many pasta pieces will be in each circle. For example, direct students to count out 24 pieces of pasta and arrange them in 6 equal groups. How many pasta pieces will be in each group? (4) Write the related division sentence.

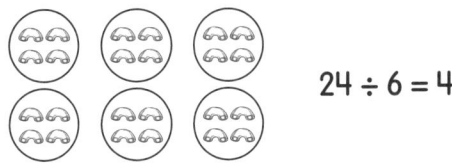

24 ÷ 6 = 4

Dividing with Beans

For this activity, provide each pair of students with a large scoop of beans in a zippered plastic bag along with a sheet of paper and a watercolor marker. Students should label the paper with quotients 1–10.

Begin the lesson by having each player reach into the bag, grab a handful of beans, and arrange them in sets to model a division problem. For example, if 8 beans were drawn, the following problems could be depicted: 8 ÷ 2 = 4, 8 ÷ 4 = 2, 8 ÷ 8 = 1, and 8 ÷ 1 = 8. Each generated math sentence is recorded on the paper near its correct quotient and is marked with an extra bean. When all possible problems have been recorded, the players return the beans to the bag and each grabs a new handful of beans. Continue in the same manner by generating at least one different math problem for each quotient until the paper is filled in.

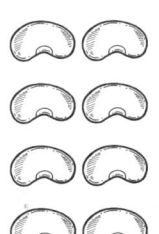
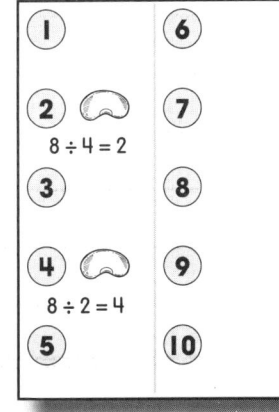

Name _____ Date _____

Directions on page 15

100's Chart

									0
1	2	3	4	5	6	7	8	9	10
11	12	13	14	15	16	17	18	19	20
21	22	23	24	25	26	27	28	29	30
31	32	33	34	35	36	37	38	39	40
41	42	43	44	45	46	47	48	49	50
51	52	53	54	55	56	57	58	59	60
61	62	63	64	65	66	67	68	69	70
71	72	73	74	75	76	77	78	79	80
81	82	83	84	85	86	87	88	89	90
91	92	93	94	95	96	97	98	99	100

KE-804077 © Key Education — Specific Skills: Division Facts Tips & Tricks

Name _____ Date _____

Division by Subtracting

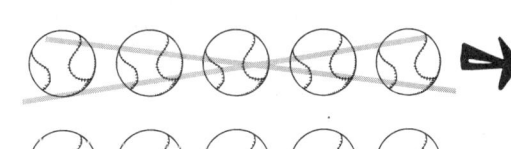

Subtracting groups of 5 balls makes 2 equal groups.

Subtract
10 − 5 = 5

5 − 5 = 0

Division Sentence
10 ÷ 5 = 2

Directions: Use repeated subtraction to find out how many equal groups can be made. Show your work. Write the related division sentence.

1.

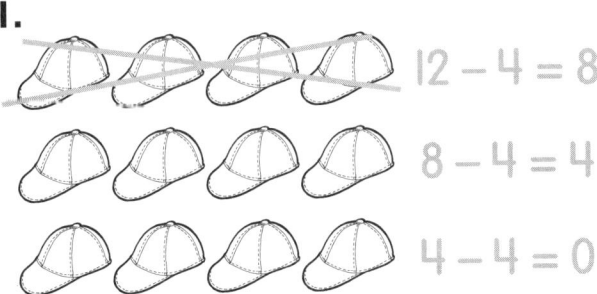

12 − 4 = 8
8 − 4 = 4
4 − 4 = 0

Subtract groups of 4.
How many groups? ____

____ ÷ ____ = ◯

2.

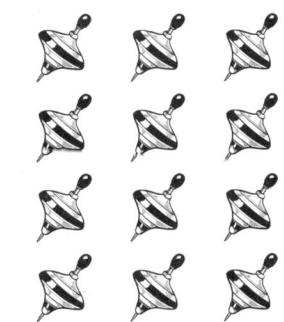

Subtract groups of 3.
How many groups? ____

____ ÷ ____ = ____

3.

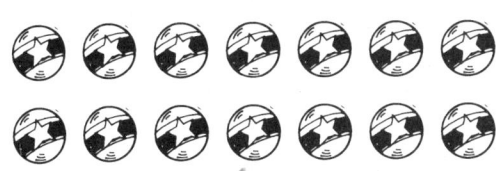

Subtract groups of 2.
How many groups? ____

____ ÷ ____ = ____

4.

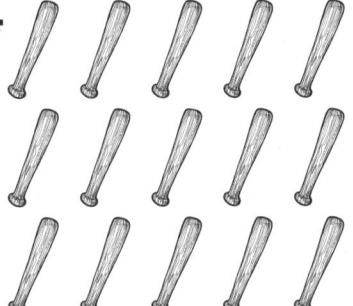

Subtract groups of 3.
How many groups? ____

____ ÷ ____ = ____

Name _____   Date _____

Subtract or Divide?

Directions: Use repeated subtraction to find out how many equal groups can be made. Write the related division sentence.

1.

18 − 6 = 12   6 − 6 = 0

12 − 6 = 6

Subtract groups of 6.
How many groups? _____

____ ÷ ____ = ◯

2.

Subtract groups of 7.
How many groups? _____

____ ÷ ____ = ____

3.

Subtract groups of 4.
How many groups? _____

____ ÷ ____ = ____

4.

Subtract groups of 8.
How many groups? _____

____ ÷ ____ = ____

KE-804077 © Key Education — 18 — Specific Skills: Division Facts Tips & Tricks

Name _____   Date _____

Making Equal Groups

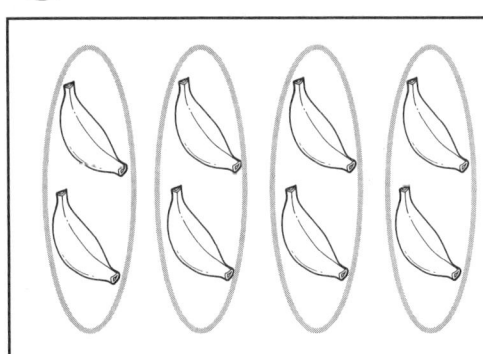

Example

Circle to make 4 equal groups.

Division Sentence

8 ÷ 4 = 2

Directions: Circle to show equal groups. Write the related division sentence.

1.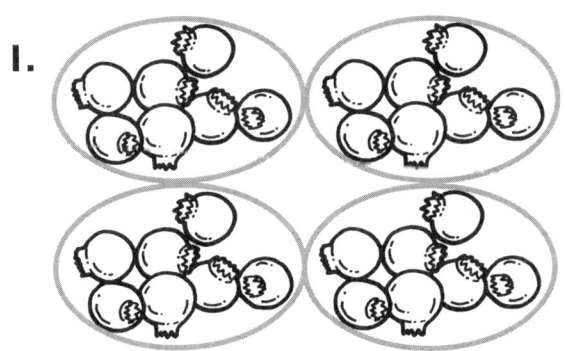

   Make 4 equal groups.
   How many in each group? ____

   ____ ÷ ____ = ____

2.

   Make 4 equal groups.
   How many in each group? ____

   ____ ÷ ____ = ____

3.

   Make 2 equal groups.
   How many in each group? ____

   ____ ÷ ____ = ____

4.

   Make 6 equal groups.
   How many in each group? ____

   ____ ÷ ____ = ____

Name _____   Date _____

Circling Equal Groups

Directions: Circle to show equal groups.
Write the related division sentence.

1.

    Make 6 equal groups.
    How many in each group? _____

    ____ ÷ ____ = ◯

2.

    Make 4 equal groups.
    How many in each group? _____

    ____ ÷ ____ = ____

3.

    Make 3 equal groups.
    How many in each group? _____

    ____ ÷ ____ = ____

4.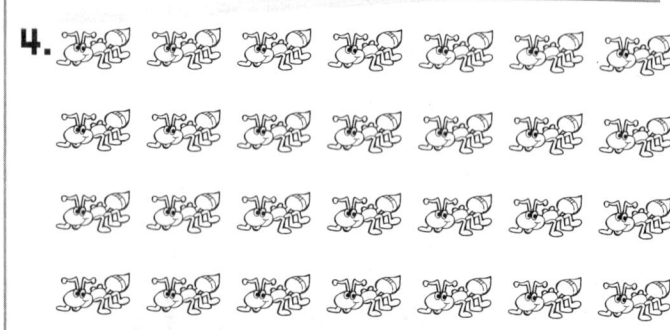

    Make 7 equal groups.
    How many in each group? _____

    ____ ÷ ____ = ____

5.

    Make 5 equal groups.
    How many in each group? _____

    ____ ÷ ____ = ____

6.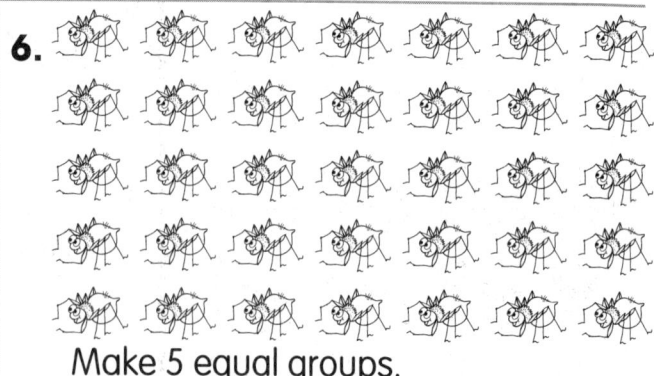

    Make 5 equal groups.
    How many in each group? _____

    ____ ÷ ____ = ____

Name _____ Date _____

Think and Draw Equal Groups

Directions: Read each problem. Draw a line to the correct pictures. Circle to show the equal groups. Write the quotient in the blank.

A.
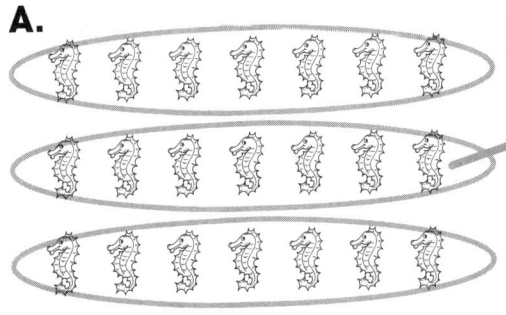

B.

C.
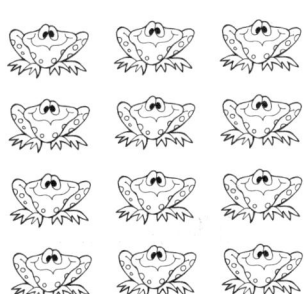

D.
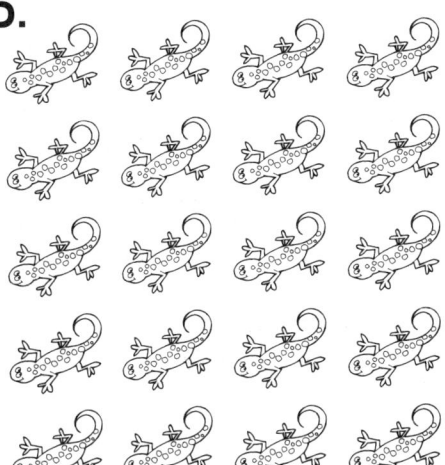

1. $21 \div 3 =$ __7__

2. $12 \div 3 =$ ____

3. $24 \div 4 =$ ____

4. $6 \div 3 =$ ____

5. $5 \div 5 =$ ____

6. $8 \div 2 =$ ____

7. $20 \div 5 =$ ____

8. $10 \div 5 =$ ____

E.
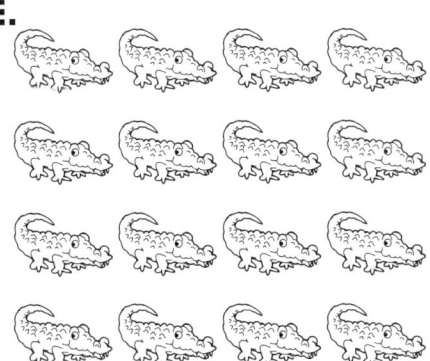

F.

G.

H.

KE-804077 © Key Education — Specific Skills: Division Facts Tips & Tricks

Name _____   Date _____

More Think and Draw Equal Groups

Directions: Read each problem. Draw the pictures. Circle to show the equal groups. Write the related division sentence.

A. Show 9 cookies.
Make 3 equal groups.

How many cookies in each group? ____   ____ ÷ ____ = ____

B. Show 18 rubber balls.
Make 6 equal groups.

How many balls in each group? ____   ____ ÷ ____ = ____

C. Show 36 candy bars.
Make 4 equal groups.

How many candy bars in each group? ____   ____ ÷ ____ = ____

D. Show 24 cell phones.
Make 3 equal groups.

How many phones in each group? ____   ____ ÷ ____ = ____

KE-804077 © Key Education — 22 — Specific Skills: Division Facts Tips & Tricks

Name _____   Date _____

Skip Counting Backwards

Directions:
- Use the chart to skip count backwards if needed.
- Write the multiples on the line.
- Tell how many skips were made.
- Write the division sentence.

									0
1	2	3	4	5	6	7	8	9	10
11	12	13	14	15	16	17	18	19	20
21	22	23	24	25	26	27	28	29	30
31	32	33	34	35	36	37	38	39	40
41	42	43	44	45	46	47	48	49	50
51	52	53	54	55	56	57	58	59	60
61	62	63	64	65	66	67	68	69	70
71	72	73	74	75	76	77	78	79	80
81	82	83	84	85	86	87	88	89	90
91	92	93	94	95	96	97	98	99	100

1. Count backwards by **5s**. Start on **50**: 45, 40, 35, 30, 25, 20, 15, 10, 5, 0

   Number of skips: ____   Division sentence: ____ ÷ 5 = ____

2. Count backwards by **6s**. Start on **48**: _____

   Number of skips: ____   Division sentence: ____ ÷ 6 = ____

3. Count backwards by **7s**. Start on **63**: _____

   Number of skips: ____   Division sentence: ____ ÷ 7 = ____

4. Count backwards by **9s**. Start on **36**: _____

   Number of skips: ____   Division sentence: ____ ÷ 9 = ____

5. Count backwards by **4s**. Start on **28**: _____

   Number of skips: ____   Division sentence: ____ ÷ 4 = ____

6. Count backwards by **8s**. Start on **64**: _____

   Number of skips: ____   Division sentence: ____ ÷ 8 = ____

KE-804077 © Key Education — Specific Skills: Division Facts Tips & Tricks

Name _____   Date _____

Skip to the Quotient

Directions:
- Read each problem.
- Use the chart to skip count backwards if needed.
- Write the quotient.
- Hint: The quotient is the number of skips.

1	2	3	4	5	6	7	8	9	0
1	2	3	4	5	6	7	8	9	10
11	12	13	14	15	16	17	18	19	20
21	22	23	24	25	26	27	28	29	30
31	32	33	34	35	36	37	38	39	40
41	42	43	44	45	46	47	48	49	50
51	52	53	54	55	56	57	58	59	60
61	62	63	64	65	66	67	68	69	70
71	72	73	74	75	76	77	78	79	80
81	82	83	84	85	86	87	88	89	90
91	92	93	94	95	96	97	98	99	100

A. 40 ÷ 5 = _____

B. 48 ÷ 8 = _____

C. 18 ÷ 9 = _____

D. 30 ÷ 5 = _____

E. 81 ÷ 9 = _____

F. 36 ÷ 6 = _____

G. 24 ÷ 4 = _____

H. 42 ÷ 6 = _____

I. 56 ÷ 7 = _____

J. 80 ÷ 10 = _____

K. 36 ÷ 9 = _____

L. 64 ÷ 8 = _____

KE-804077 © Key Education — 24 — *Specific Skills: Division Facts Tips & Tricks*

Name _____ Date _____

Picture the Fact Families

Multiplication Sentences	Division Sentences
4 × 3 = 12	12 ÷ 4 = 3
3 × 4 = 12	12 ÷ 3 = 4

This array shows the fact family for 3, 4, and 12.

Directions: Look at each array. Count the rows and columns to write two multiplication sentences and two division sentences.

1.

___ × ___ = ___
___ × ___ = ___
___ ÷ ___ = ___
___ ÷ ___ = ___

2.

___ × ___ = ___
___ × ___ = ___
___ ÷ ___ = ___
___ ÷ ___ = ___

3.

___ × ___ = ___
___ × ___ = ___
___ ÷ ___ = ___
___ ÷ ___ = ___

4.

___ × ___ = ___
___ × ___ = ___
___ ÷ ___ = ___
___ ÷ ___ = ___

Name _____   Date _____

Smart Thinking About Fact Families

| Set of Numbers 2, 4, 8 | Multiplication Sentences 2 x 4 = 8 4 x 2 = 8 | Division Sentences 8 ÷ 2 = 4 8 ÷ 4 = 2 |

Directions: Write the fact family for each set of numbers.

A. 6, 8, 48

B. 6, 5, 30

C. 3, 5, 15

D. 5, 8, 40

E. 6, 7, 42

F. 7, 4, 28

G. 8, 3, 24

H. 9, 2, 18

I. 3, 9, 27

J. 4, 9, 36

KE-804077 © Key Education — Specific Skills: Division Facts Tips & Tricks

Name _____   Date _____

Matching Fact Equals Matching Fact

Division facts can be written two ways.
Example: $16 \div 8$ and $8\overline{)16}$

Directions: Draw a line to match the division fact in column A with the correct quotient in column B. Then, draw a line to the correct division fact in column C.

Column A	Column B	Column C
A. $12 \div 6 =$	7	$16 \div 4 =$
B. $4\overline{)16}$	8	$5\overline{)25}$
C. $25 \div 5 =$	5	$18 \div 3 =$
D. $4\overline{)28}$	4	$4\overline{)32}$
E. $32 \div 4 =$	3	$6\overline{)12}$
F. $3\overline{)18}$	9	$28 \div 4 =$
G. $7 \div 7 =$	6	$7\overline{)7}$
H. $9 \div 3 =$	2	$40 \div 4 =$
I. $4\overline{)40}$	1	$4\overline{)36}$
J. $36 \div 4 =$	10	$3\overline{)9}$

KE-804077 © Key Education — 27 — Specific Skills: Division Facts Tips & Tricks

Name _____ Date _____

Check the Division Facts

Directions: Look at the division facts below. Circle the facts that are correct.

A. 6 ÷ 6 = 6	**B.** 24 ÷ 6 = 4	**C.** 63 ÷ 7 = 9
D. 15 ÷ 3 = 5	**E.** 27 ÷ 3 = 8	**F.** 30 ÷ 5 = 7
G. 16 ÷ 4 = 4	**H.** 45 ÷ 5 = 8	**I.** 81 ÷ 9 = 9
J. 18 ÷ 6 = 4	**K.** 56 ÷ 8 = 7	**L.** 72 ÷ 9 = 8

Directions: Copy each incorrect division fact shown above and correct the answer.

_____   _____

_____   _____

_____   _____

Directions: Find the quotient.

M. 3)21   **N.** 6)42

O. 14 ÷ 7 = ____   **P.** 20 ÷ 5 =   **Q.** 63 ÷ 9 = ____

R. 6)30   **S.** 40 ÷ 8 = ____   **T.** 4)32

U. 6)36   **V.** 7)49   **W.** 8)64

KE-804077 © Key Education   28   Specific Skills: Division Facts Tips & Tricks

Name _____   Date _____

A Trick for Even-Number Divisors

Halve It, Double It

- Look at the divisor:
  Half of 4 is 2.

- Double the quotient:
  3 + 3 = 6

$12 \div 4 = 3$
↓   ↓
$12 \div 2 = 6$

Use this trick to learn more facts quickly!

Directions: For each given fact, follow the directions. Write the new fact.

A. $8 \div 4 = 2$

Halve the divisor: ____

Double the quotient: ____

New fact:

$8 \div \underline{} = \underline{}$

B. $16 \div 8 = 2$

Halve the divisor: ____

Double the quotient: ____

New fact:

$16 \div \underline{} = \underline{}$

C. $24 \div 6 = 4$

Halve the divisor: ____

Double the quotient: ____

New fact:

$24 \div \underline{} = \underline{}$

D. $32 \div 8 = 4$

Halve the divisor: ____

Double the quotient: ____

New fact:

$32 \div \underline{} = \underline{}$

E. $20 \div 4 = 5$

Halve the divisor: ____

Double the quotient: ____

New fact:

$20 \div \underline{} = \underline{}$

F. $14 \div 2 = 7$

Halve the divisor: ____

Double the quotient: ____

New fact:

$14 \div \underline{} = \underline{}$

Think About This! Dividing Evens and Odds

Directions: Solve. Circle each even quotient in red. Circle each odd quotient in black.

A. $15 \div 3 =$ ____   **B.** $24 \div 6 =$ ____   **C.** $27 \div 3 =$ ____   **D.** $18 \div 9 =$ ____

E. $12 \div 2 =$ ____   **F.** $4 \div 2 =$ ____   **G.** $16 \div 4 =$ ____

H. $7 \div 7 =$ ____   **I.** $10 \div 5 =$ ____   **J.** $20 \div 5 =$ ____

Try other facts! Circle the patterns below that always work.

Look at each problem above. Write the word "even" or "odd" on the line if the pattern works.

K. even ÷ even = _____   **L.** odd ÷ odd = _____   **M.** even ÷ odd = _____

Name _____   Date _____

Dividing by 0 and 1

- **Numbers cannot be divided by 0 (zero).** It is not **possible** to make 0 (zero) equal groups. For example: 3 ÷ 0 is meaningless.
- **Zero divided by any number is 0 (zero).** For example: 0 ÷ 3 = 0
- **Any number divided by 1 (one) is that number.** There will always be 1 (one) group and everything is in that group. For example: 3 ÷ 1 = 3

Directions: Write the quotient. If you need help, draw pictures. Circle to show equal groups.

A. 4 ÷ 1 = ____

B. 0 ÷ 2 = ____

C. 9 ÷ 1 = ____

D. 1 ÷ 1 = ____

E. 0 ÷ 7 = ____

F. 12 ÷ 1 = ____

G. 0 ÷ 6 = ____

H. 6 ÷ 1 = ____

I. 1)‾5   J. 1)‾10   K. 1)‾3   L. 5)‾0

M. 1)‾7   N. 1)‾8   O. 4)‾0   P. 1)‾9

Q. 3)‾0   R. 9)‾0   S. 1)‾0   T. 8)‾0

KE-804077 © Key Education — 30 — Specific Skills: Division Facts Tips & Tricks

Name _____   Date _____

Dividing by 2 and 5

 **To divide by 2**, simply divide the given number in half. You are putting the **dividend** into 2 equal groups.

Directions: Write the quotient. If you need help, use the pictures.

A. 8 ÷ 2 = ____   ☆ ☆ ☆ ☆ ☆ ☆ ☆ ☆

B. 10 ÷ 2 = ____   ✵ ✵ ✵ ✵ ✵ ✵ ✵ ✵ ✵ ✵

C. 18 ÷ 2 = ____   ▲▲▲▲▲▲▲▲▲▲▲▲▲▲▲▲▲▲

D. 14 ÷ 2 = ____   O O O O O O O O O O O O O O

E. 12 ÷ 2 = ____   ★ ★ ★ ★ ★ ★ ★ ★ ★ ★ ★ ★

F. 16 ÷ 2 = ____   ☐☐☐☐☐☐☐☐☐☐☐☐☐☐☐☐

 **To divide by 5**, simply count by 5s until you reach the **dividend**. The number of skip counts is the **quotient**.

Directions: Write the quotient. If you need help, count by 5s.

G. 20 ÷ 5 = ____   **H.** 50 ÷ 5 = ____

I. 35 ÷ 5 = ____   **J.** 45 ÷ 5 = ____

K. 30 ÷ 5 = ____   **L.** 40 ÷ 5 = ____

KE-804077 © Key Education — Specific Skills: Division Facts Tips & Tricks

Name _____ Date _____

More Dividing by 2 and 5

- **To divide by 2**, simply divide the given number in half. You are putting the **dividend** into 2 equal groups. Any **even** number can be divided by 2.

- **To divide by 5**, simply count by 5s until you reach the **dividend**. The number of skip counts is the **quotient**.

Directions: Write the quotient.

A. 2 ÷ 2 = ____

B. 10 ÷ 5 = ____

C. 4 ÷ 2 = ____

D. 25 ÷ 5 = ____

E. 14 ÷ 2 = ____

F. 10 ÷ 2 = ____

G. 35 ÷ 5 = ____

H. 16 ÷ 2 = ____

I. 15 ÷ 5 = ____

J. 18 ÷ 2 = ____

K. 6 ÷ 2 = ____

L. 20 ÷ 5 = ____

M. 12 ÷ 2 = ____

N. 50 ÷ 5 = ____

O. 20 ÷ 2 = ____

Name _____   Date _____

Divisible by Doubles

Divisible by doubles involves facts that have identical divisors and quotients.

Example: 9 ÷ 3 = 3

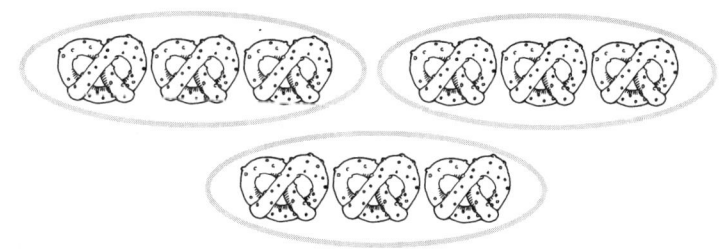

Directions: Use tally marks or draw pictures to show the answer. Write the quotient.

A. 16 ÷ 4 = ____

| I I I I | I I I I |
| I I I I | I I I I |

B. 25 ÷ 5 = ____

C. 36 ÷ 6 = ____

D. 64 ÷ 8 = ____

E. 81 ÷ 9 = ____

F. 4 ÷ 2 = ____

G. 9 ÷ 3 = ____

H. 49 ÷ 7 = ____

I. 1 ÷ 1 = ____

Name _____ Date _____

Crossword Quotients

Directions: Find the quotient. Write the answer as a word in the puzzle.

ACROSS
3. 36 ÷ 6 = _____
4. 1 ÷ 1 = _____
6. 25 ÷ 5 = _____
8. 9 ÷ 3 = _____
9. 100 ÷ 10 = _____

DOWN
1. 4 ÷ 2 = _____
2. 81 ÷ 9 = _____
5. 64 ÷ 8 = _____
6. 16 ÷ 4 = _____
7. 49 ÷ 7 = _____

KE-804077 © Key Education — Specific Skills: Division Facts Tips & Tricks

Name _____ Date _____

Picture These Quotients

Directions: Write the quotient. Use the pictures below to solve each problem.

A. 12 ÷ 3 = ____

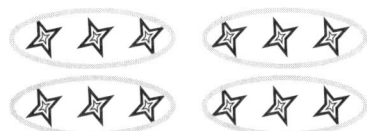

B. 9 ÷ 3 = ____

C. 24 ÷ 3 = ____

D. 15 ÷ 3 = ____

E. 18 ÷ 3 = ____

F. 27 ÷ 3 = ____

G. 21 ÷ 3 = ____

H. 20 ÷ 4 = ____

I. 32 ÷ 4 = ____

J. 12 ÷ 4 = ____

K. 36 ÷ 4 = ____

L. 16 ÷ 4 = ____

M. 28 ÷ 4 = ____

N. 24 ÷ 4 = ____

KE-804077 © Key Education — 35 — Specific Skills: Division Facts Tips & Tricks

Name _____ Date _____

Swimming with Facts

Directions: Write the quotient.

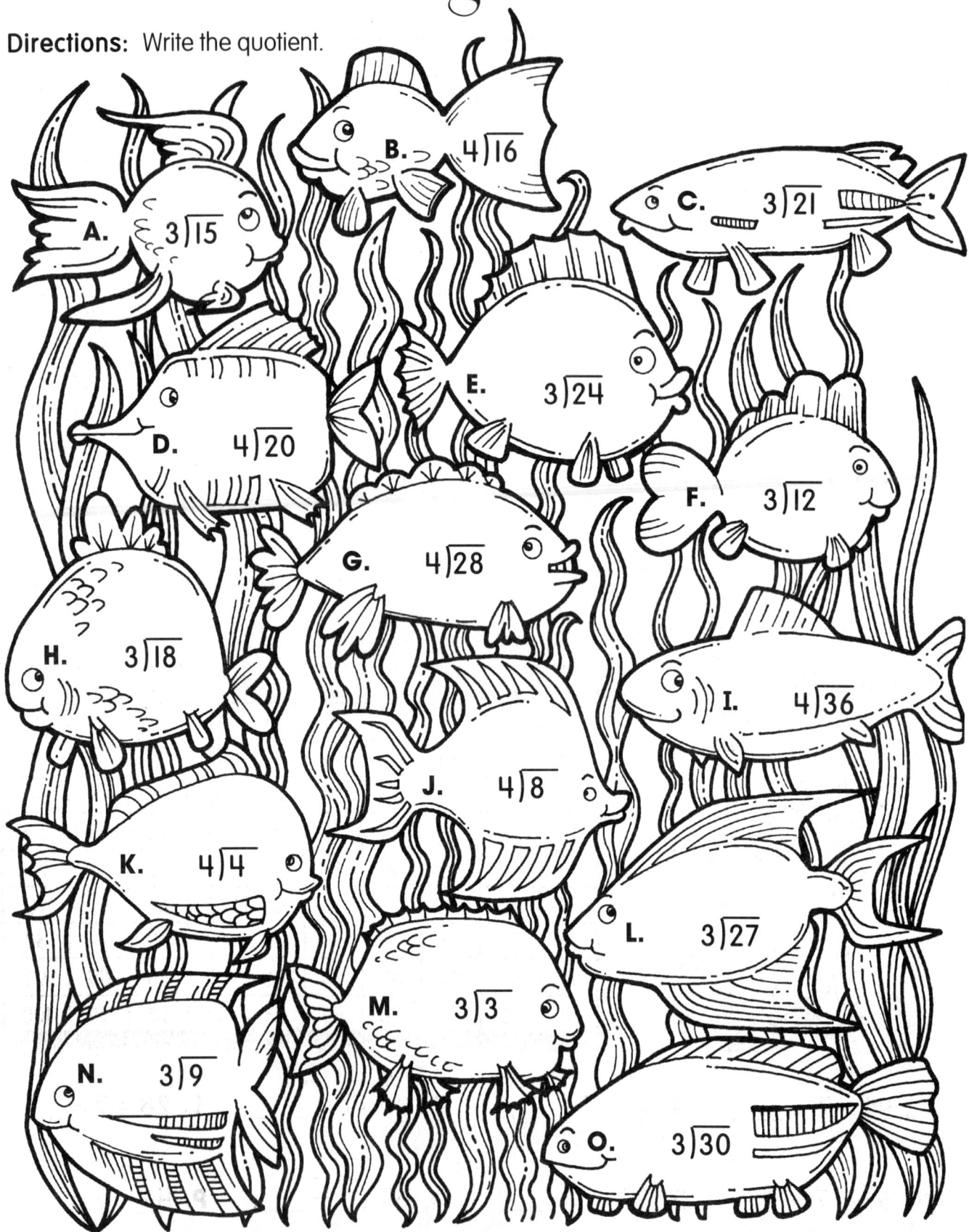

A. 3)15
B. 4)16
C. 3)21
D. 4)20
E. 3)24
F. 3)12
G. 4)28
H. 3)18
I. 4)36
J. 4)8
K. 4)4
L. 3)27
M. 3)3
N. 3)9
O. 3)30

Name _____   Date _____

Shade In and Solve

Directions: Count by 6s up to 60. Lightly shade in the boxes. Solve the problems.

1	2	3	4	5	6	7	8	9	10
11	12	13	14	15	16	17	18	19	20
21	22	23	24	25	26	27	28	29	30
31	32	33	34	35	36	37	38	39	40
41	42	43	44	45	46	47	48	49	50
51	52	53	54	55	56	57	58	59	60
61	62	63	64	65	66	67	68	69	70
71	72	73	74	75	76	77	78	79	80
81	82	83	84	85	86	87	88	89	90
91	92	93	94	95	96	97	98	99	100

A. 36 ÷ 6 = ____   **B.** 12 ÷ 6 = ____   **C.** 18 ÷ 6 = ____   **D.** 48 ÷ 6 = ____

E. 42 ÷ 6 = ____   **F.** 54 ÷ 6 = ____   **G.** 24 ÷ 6 = ____   **H.** 30 ÷ 6 = ____

Directions: Count by 7s up to 70. Lightly shade in the boxes. Solve the problems.

1	2	3	4	5	6	7	8	9	10
11	12	13	14	15	16	17	18	19	20
21	22	23	24	25	26	27	28	29	30
31	32	33	34	35	36	37	38	39	40
41	42	43	44	45	46	47	48	49	50
51	52	53	54	55	56	57	58	59	60
61	62	63	64	65	66	67	68	69	70
71	72	73	74	75	76	77	78	79	80
81	82	83	84	85	86	87	88	89	90
91	92	93	94	95	96	97	98	99	100

I. 42 ÷ 7 = ____   **J.** 21 ÷ 7 = ____   **K.** 14 ÷ 7 = ____   **L.** 28 ÷ 7 = ____

M. 56 ÷ 7 = ____   **N.** 35 ÷ 7 = ____   **O.** 63 ÷ 7 = ____   **P.** 49 ÷ 7 = ____

KE-804077 © Key Education — Specific Skills: Division Facts Tips & Tricks

Name _____   Date _____

More Shade In and Solve

Directions: Count by 8s up to 80. Lightly shade in the boxes. Solve the problems.

1	2	3	4	5	6	7	8	9	10
11	12	13	14	15	16	17	18	19	20
21	22	23	24	25	26	27	28	29	30
31	32	33	34	35	36	37	38	39	40
41	42	43	44	45	46	47	48	49	50
51	52	53	54	55	56	57	58	59	60
61	62	63	64	65	66	67	68	69	70
71	72	73	74	75	76	77	78	79	80
81	82	83	84	85	86	87	88	89	90
91	92	93	94	95	96	97	98	99	100

A. 64 ÷ 8 = ____   **B.** 16 ÷ 8 = ____   **C.** 48 ÷ 8 = ____   **D.** 56 ÷ 8 = ____

E. 40 ÷ 8 = ____   **F.** 24 ÷ 8 = ____   **G.** 72 ÷ 8 = ____   **H.** 32 ÷ 8 = ____

Directions: Count by 9s up to 90. Lightly shade in the boxes. Solve the problems.

1	2	3	4	5	6	7	8	9	10
11	12	13	14	15	16	17	18	19	20
21	22	23	24	25	26	27	28	29	30
31	32	33	34	35	36	37	38	39	40
41	42	43	44	45	46	47	48	49	50
51	52	53	54	55	56	57	58	59	60
61	62	63	64	65	66	67	68	69	70
71	72	73	74	75	76	77	78	79	80
81	82	83	84	85	86	87	88	89	90
91	92	93	94	95	96	97	98	99	100

I. 18 ÷ 9 = ____   **J.** 81 ÷ 9 = ____   **K.** 72 ÷ 9 = ____   **L.** 36 ÷ 9 = ____

M. 63 ÷ 9 = ____   **N.** 45 ÷ 9 = ____   **O.** 27 ÷ 9 = ____   **P.** 54 ÷ 9 = ____

Name _____   Date _____

Hit the Mark!

Directions: Divide. Write the number of correct answers for each row in the box.

Score:

A.

| 6)12 | 6)30 | 6)48 | 7)21 | 7)49 | 8)48 | 7)70 | 9)90 |

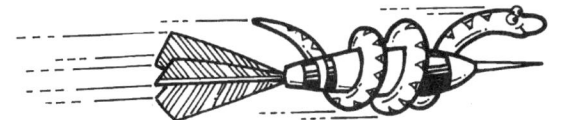 

Score:

B.

| 9)45 | 7)63 | 7)35 | 6)60 | 6)36 | 8)56 | 8)72 | 6)18 |

Score:

C.

| 6)0 | 7)14 | 7)28 | 9)18 | 9)36 | 6)54 | 6)42 | 7)7 |

Score:

D.

| 9)54 | 8)16 | 7)42 | 6)6 | 7)0 | 8)24 | 9)81 | 9)72 |

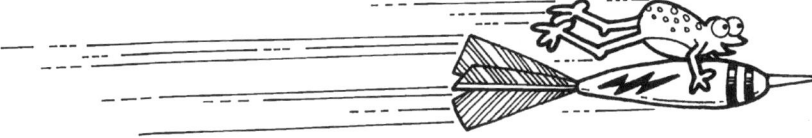 

Score:

E.

| 6)24 | 7)56 | 9)27 | 8)32 | 8)40 | 9)63 | 9)54 | 8)64 |

KE-804077 © Key Education   39   Specific Skills: Division Facts Tips & Tricks

Name _____ Date _____

Read and Solve

Directions: Read each problem and find the quotient. Show your work.

1. Kelly has 12 kittens. She has 6 friends who want kittens. How many kittens will each friend receive?

   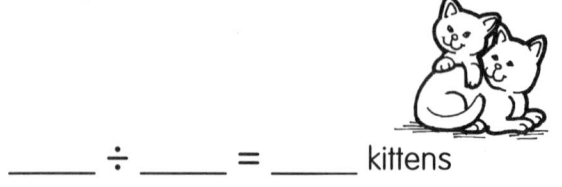

   ____ ÷ ____ = ____ kittens

2. Lauren has made 32 place mats to sell. If she puts them in stacks of 8, how many stacks will she make?

   ____ ÷ ____ = ____ stacks

3. Debbie set up the art supplies with 27 bottles of paint. She arranges the supplies into groups of 3 bottles. How many groups does she make?

   ____ ÷ ____ = ____ groups

4. Clint received a bag of jelly beans. There are 72 jelly beans in the bag. He wants to divide them equally among 9 friends. How many jelly beans will each friend receive?

   ____ ÷ ____ = ____ jelly beans

5. Jennifer purchased 18 books as gifts for her friends. If she gives each friend 3 books in a box, how many boxes will she need to wrap?

   ____ ÷ ____ = ____ boxes

6. Christian plays a horn in the marching band. There are 54 members in the band with 6 students in each row. How many rows can they form?

   ____ ÷ ____ = ____ rows

7. Dan bought 42 pencil top erasers for his 7 friends. How many erasers will each friend receive?

   ____ ÷ ____ = ____ erasers

8. Josh's dad filled the tank of his car with 8 gallons of gas. He can drive 80 miles before refilling his gas tank. How many miles can he drive on each gallon of gas?

   ____ ÷ ____ = ____ miles

Name _____   Date _____

Smart and Easy Dividing by 10

Dividing by 10 is easy when using this trick! Simply erase one 0 (zero) from the dividend.

For example: What is 30 ÷ 10?

To find out, write the dividend (30) on paper. Then, erase the zero. What is left? The 3 is left, so **30 ÷ 10 = 3**.

Directions: Solve to find the quotient.

A. 20 ÷ 10 = ____   B. 60 ÷ 10 = ____   C. 30 ÷ 10 = ____   D. 80 ÷ 10 = ____

E. 40 ÷ 10 = ____   F. 100 ÷ 10 = ____   G. 70 ÷ 10 = ____   H. 50 ÷ 10 = ____

I. 10 ÷ 10 = ____   J. 0 ÷ 10 = ____   K. 90 ÷ 10 = ____

Directions: Fill in the missing dividend, divisor, or quotient.

L. 10)‾100   M. ___)‾10 (1)   N. ___)‾30 (3)   O. 6)‾___ (10)

P. 5)‾50   Q. 2)‾20   R. ___)‾90 (9)   S. 8)‾___ (10)

T. 7)‾70   U. ___)‾40 (10)   V. ___)‾80 (8)

W. There are 80 students going on the science field trip. Each of the 8 teachers will have an equal number of students in a group. How many students will be in each group? ____ students

Show your work.   ____ ÷ ____ = ____

KE-804077 © Key Education — 41 — Specific Skills: Division Facts Tips & Tricks

Name _____   Date _____

Tic-Tac-Divide 1, 2, 5

Directions: Solve the problems. To find out the winner of each game, draw an **O** on each problem with an odd quotient. Draw an **X** on each problem with an even quotient.

1)5	5)30	2)10
2)2	2)16	5)35
2)20	5)45	1)9

2)8	5)15	2)6
2)18	5)40	5)20
5)35	2)12	2)4

2)14	1)6	5)5
5)10	1)1	1)8
1)7	5)25	5)50

KE-804077 © Key Education — 42 — Specific Skills: Division Facts Tips & Tricks

Name _____ Date _____

Leapin' Lily Pad Math

Directions: Solve the problems. Show how each frog leaps from lily pad to lily pad. Draw a blue path to connect all odd number answers. Then, draw a green path to connect all even number answers. **Hint:** One frog jumps over one of the other frog's lily pads.

KE-804077 © Key Education — 43 — Specific Skills: Division Facts Tips & Tricks

Name _____ Date _____

Climb to the Top

Directions: Write the quotient below each problem.

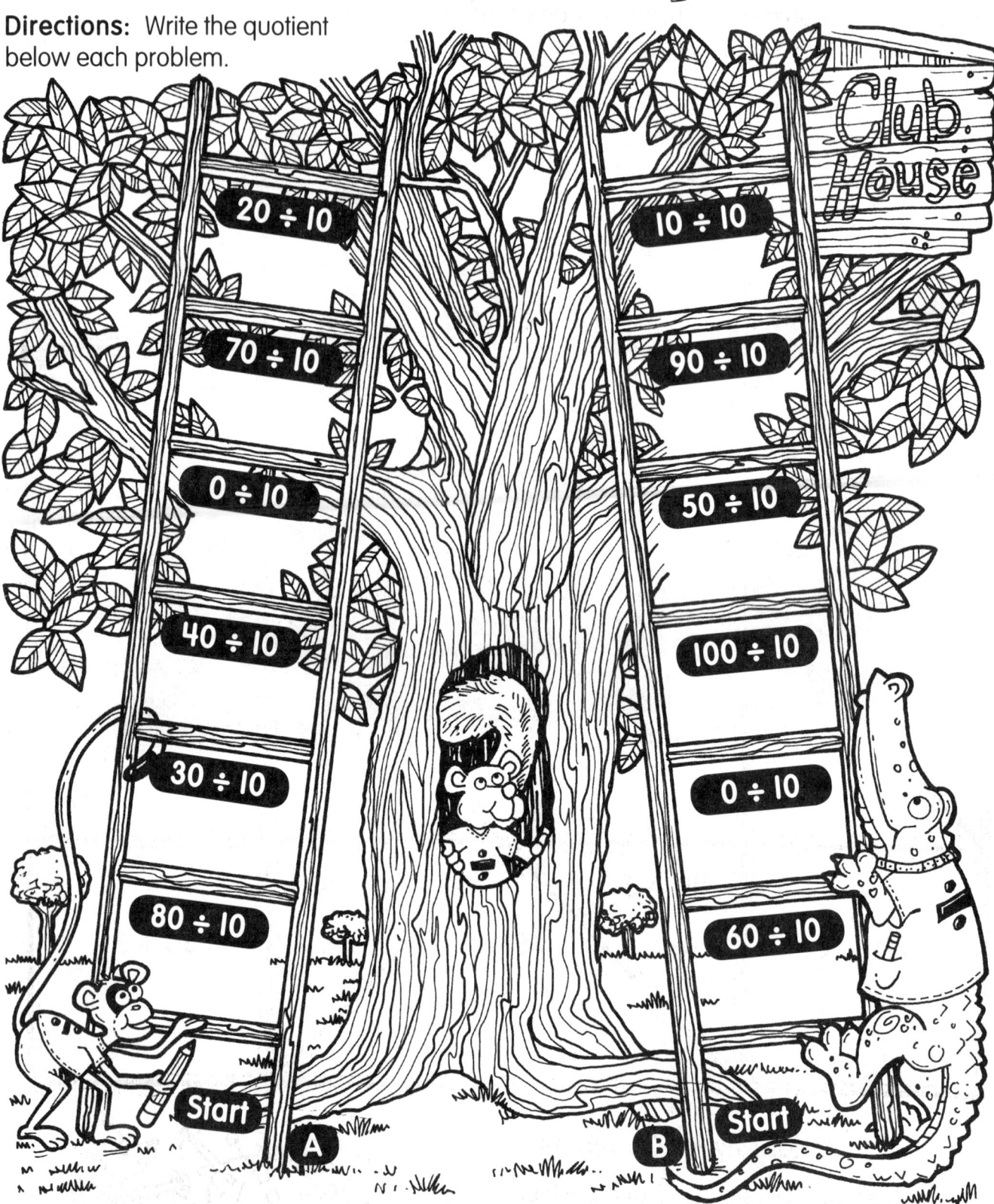

Something to Try! Use the "Dividing by 10" trick on larger quotients, such as 200, 300, 540, and so on. Write the problems on the back of this paper and have a friend solve them.

KE-804077 © Key Education — 44 — Specific Skills: Division Facts Tips & Tricks

Name _____   Date _____

Ohio Star Quilt Block

Directions: The Ohio star quilt block is an American quilt pattern. Solve each problem and write the quotient below it. Then, color each section using the color key to finish this modified quilt block.

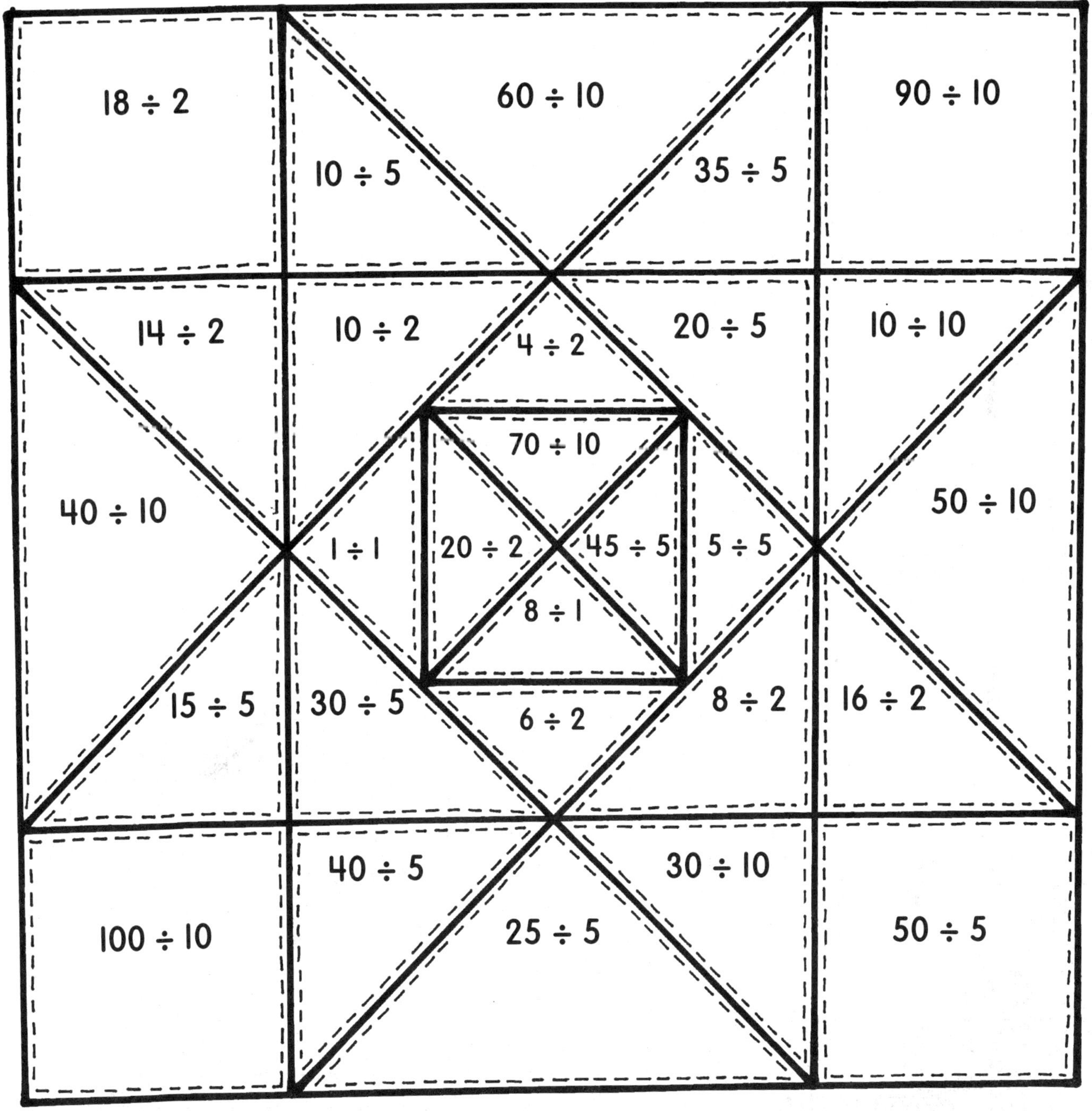

18 ÷ 2	60 ÷ 10	90 ÷ 10
	10 ÷ 5 35 ÷ 5	
14 ÷ 2	10 ÷ 2 4 ÷ 2 20 ÷ 5	10 ÷ 10
40 ÷ 10	70 ÷ 10	50 ÷ 10
	1 ÷ 1 20 ÷ 2 45 ÷ 5 5 ÷ 5	
	8 ÷ 1	
	15 ÷ 5 30 ÷ 5 6 ÷ 2 8 ÷ 2 16 ÷ 2	
	40 ÷ 5 30 ÷ 10	
100 ÷ 10	25 ÷ 5	50 ÷ 5

Color Key for Quotients

1, 2, and 3 = Green   7 and 8 = Blue
4, 5, and 6 = Red   9 and 10 = Yellow

KE-804077 © Key Education   45   Specific Skills: Division Facts Tips & Tricks

Name _____ Date _____

On Your Mark, Get Set, Divide!

Directions: Solve each problem and write the quotient.

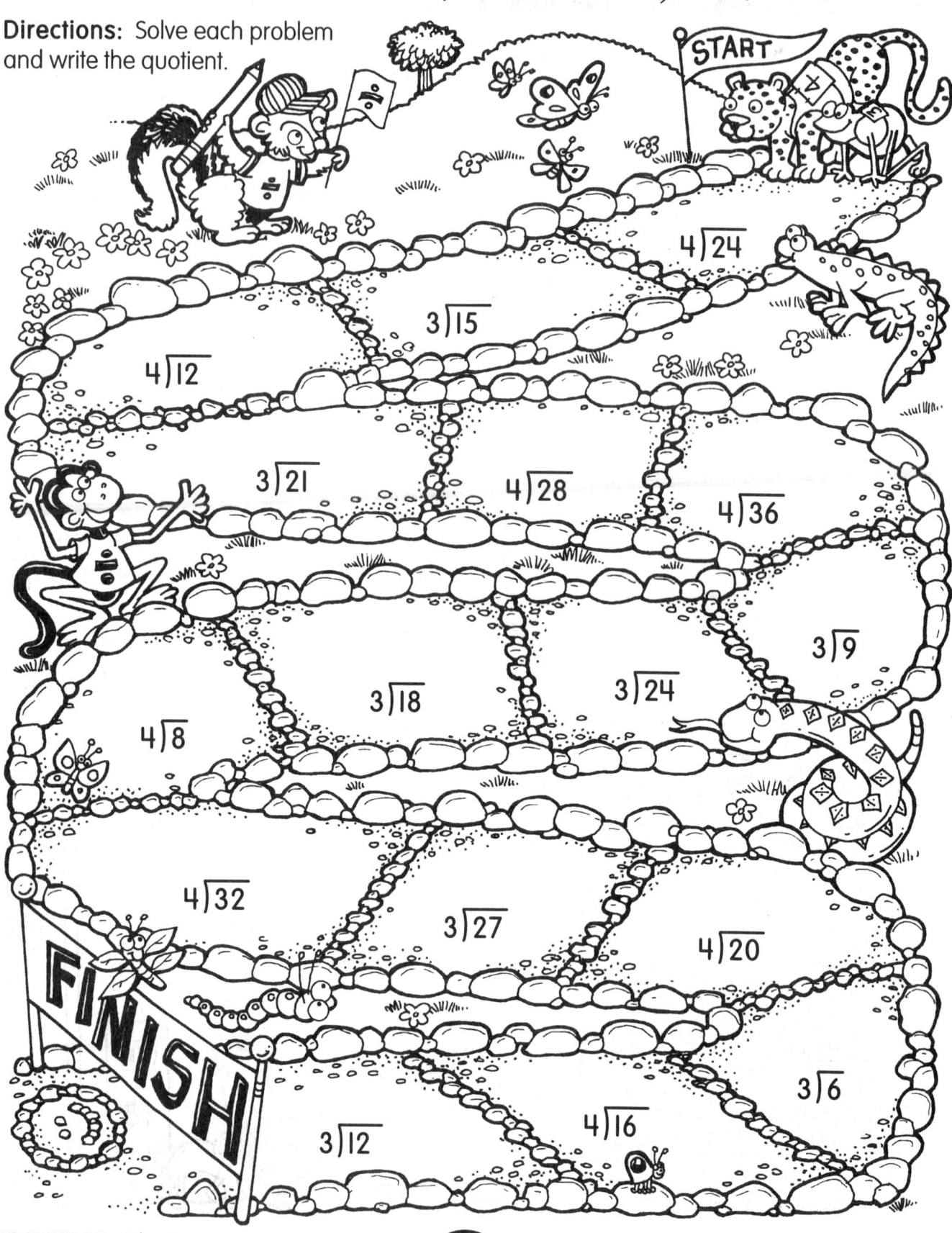

4)24
3)15
4)12
3)21
4)28
4)36
4)8
3)18
3)24
3)9
4)32
3)27
4)20
3)6
3)12
4)16

Solve the Case of the Missing Numbers

Directions: Draw a line to match each division fact with its dividend. Then, write the number in the blank.

A. _____ ÷ 3 = 7

B. _____ ÷ 4 = 6

C. _____ ÷ 3 = 4

D. _____ ÷ 4 = 10

E. _____ ÷ 3 = 9

F. _____ ÷ 4 = 4

G. _____ ÷ 3 = 10

H. _____ ÷ 4 = 8

I. _____ ÷ 3 = 6

J. _____ ÷ 4 = 9

Name _____ Date _____

Lift Off with the Facts!

Directions: Solve each problem and fill in the missing number.

A
- 24 ÷ 6 = ___
- ___ ÷ 6 = 8
- 42 ÷ 6 = ___
- ___ ÷ 6 = 10
- 30 ÷ 6 = ___
- ___ ÷ 6 = 3
- 36 ÷ 6 = ___
- 6 ÷ 6 = ___
- ___ ÷ 6 = 2
- 54 ÷ 6 = ___

B
- 63 ÷ 9 = ___
- ___ ÷ 9 = 2
- 36 ÷ 9 = ___
- 90 ÷ 9 = ___
- ___ ÷ 9 = 5
- ___ ÷ 9 = 1
- 72 ÷ 9 = ___
- ___ ÷ 9 = 9
- 27 ÷ 9 = ___
- ___ ÷ 9 = 6

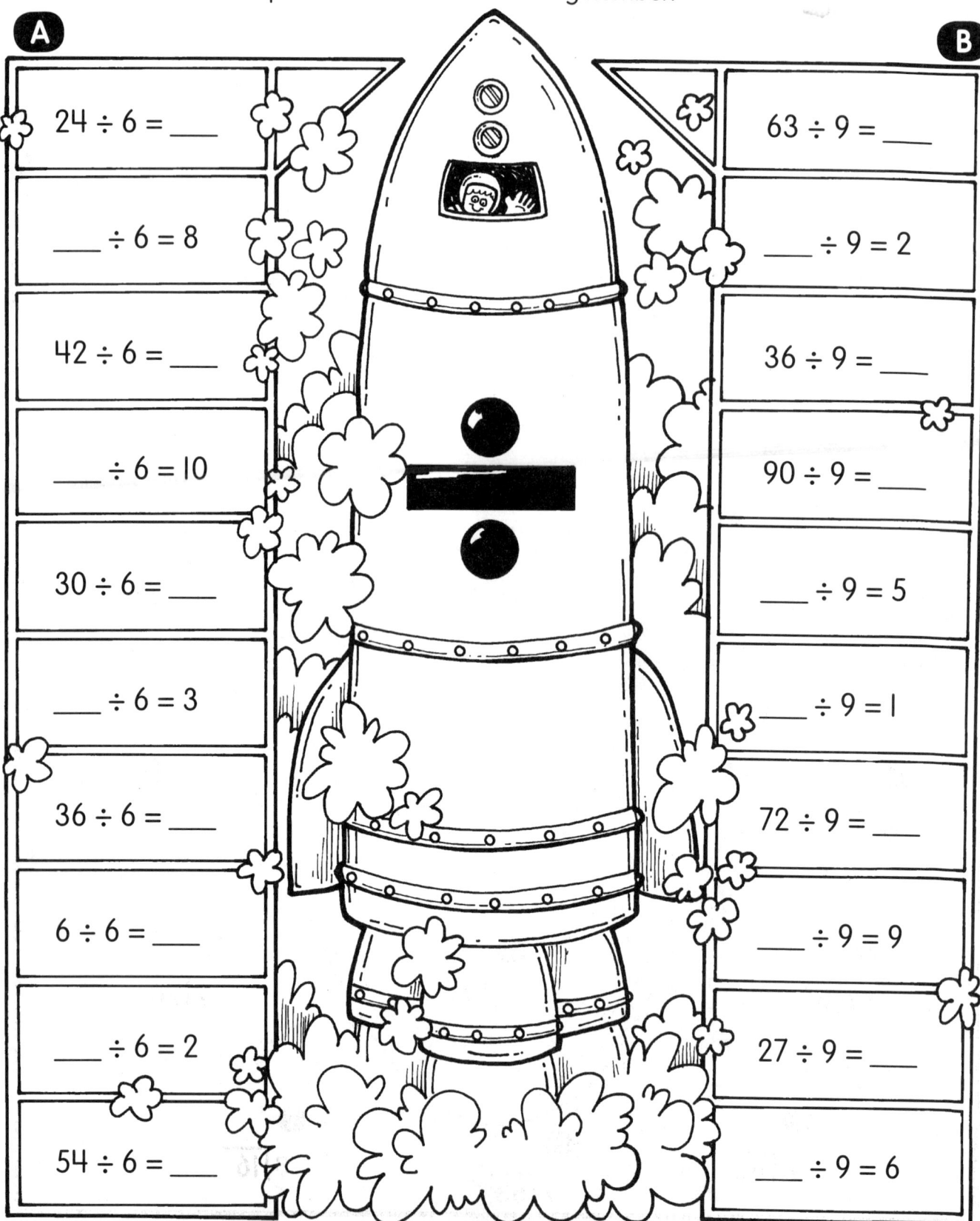

KE-804077 © Key Education — Specific Skills: Division Facts Tips & Tricks

Catch Air with Facts

Directions: Solve each problem and write the quotient.

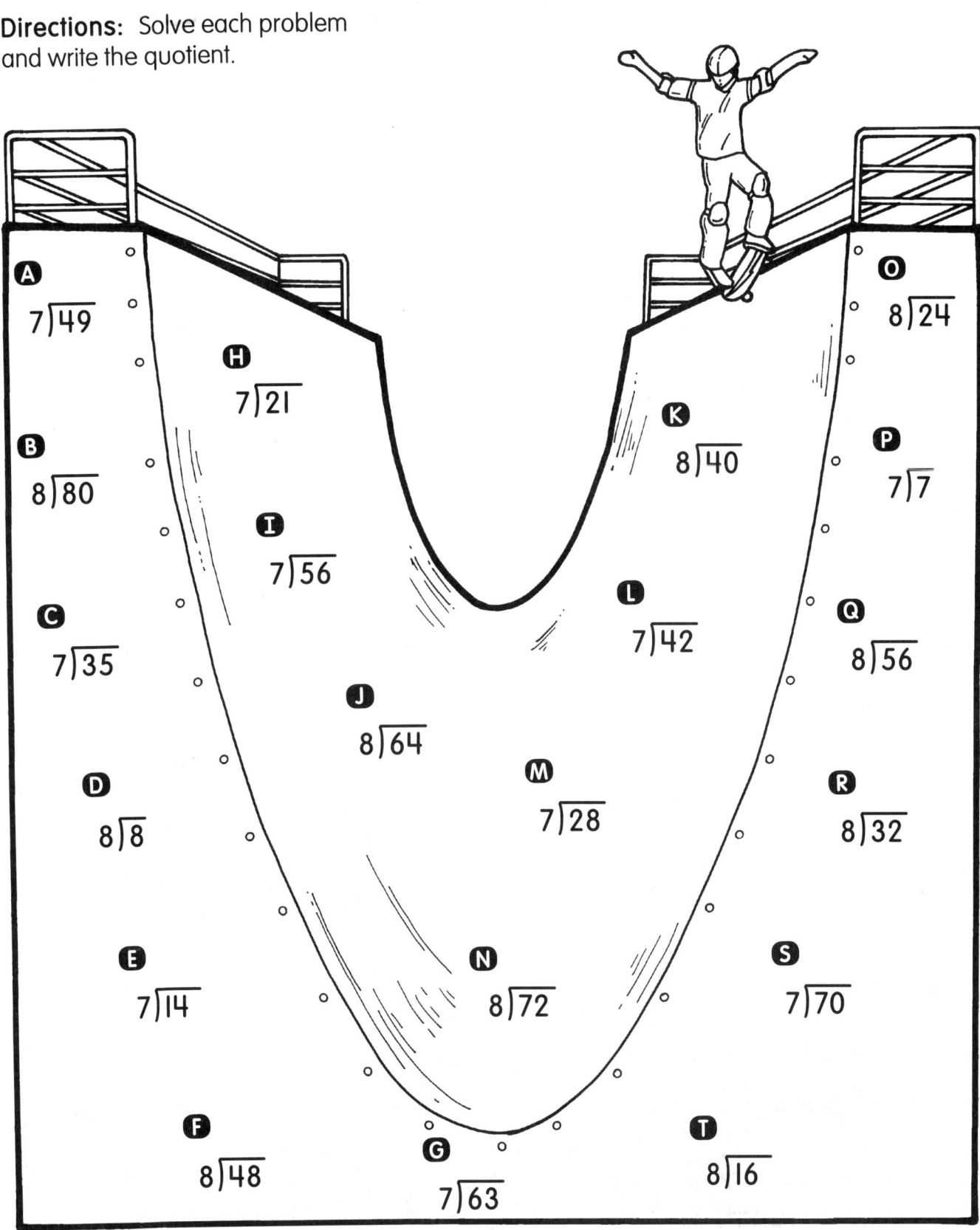

- A. 7)49
- B. 8)80
- C. 7)35
- D. 8)8
- E. 7)14
- F. 8)48
- G. 7)63
- H. 7)21
- I. 7)56
- J. 8)64
- K. 8)40
- L. 7)42
- M. 7)28
- N. 8)72
- O. 8)24
- P. 7)7
- Q. 8)56
- R. 8)32
- S. 7)70
- T. 8)16

Name _____ Date _____

Rev Up for Division

Directions: Solve each problem and write the quotient in the outer ring.

Number Correct: _____

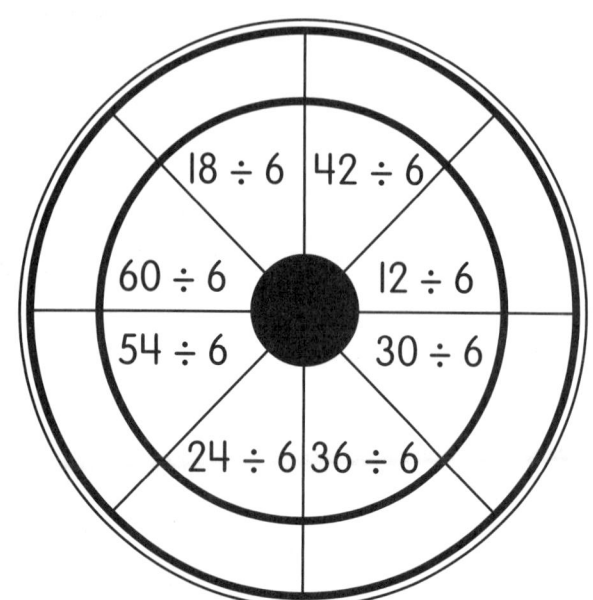
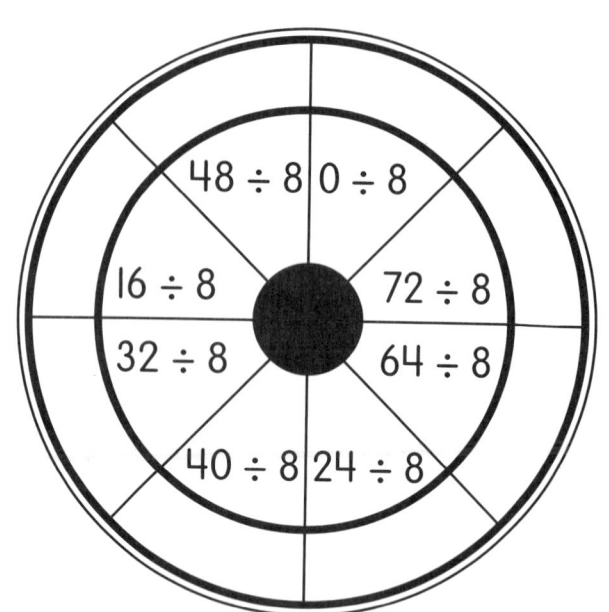

Number Correct: _____

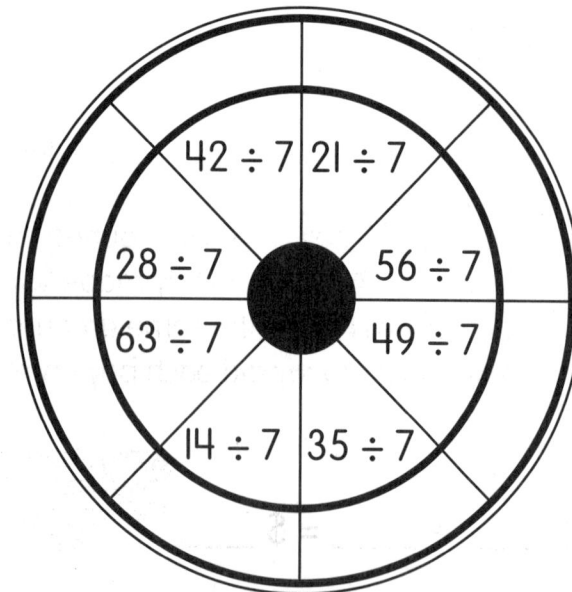
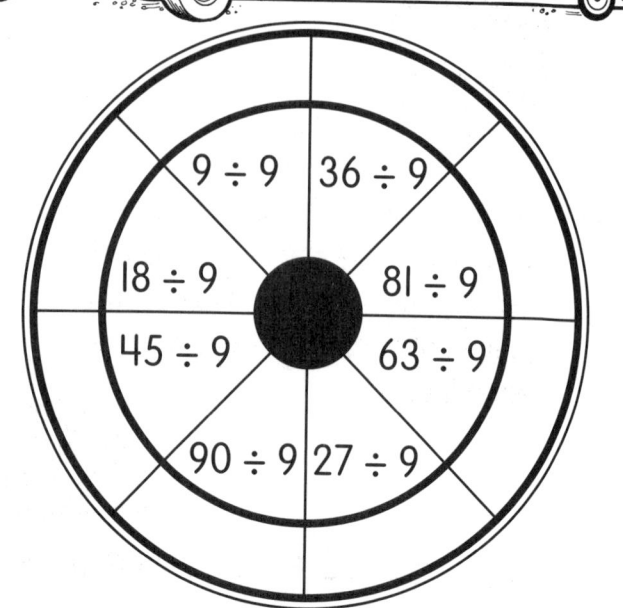

KE-804077 © Key Education — 50 — Specific Skills: Division Facts Tips & Tricks

Name _____   Date _____

More Read and Solve

Directions: Read each problem and find the quotient. Show your work.

1. There are 21 students in Mrs. McCoy's class. She divides the students into 7 equal groups. How many students are in each group?

    ____ ÷ ____ = ____ students

2. Ms. Branch bought 9 identical beach towels on sale for $81. How much did she pay for each towel?

    $ ____ ÷ ____ = $ ____

3. For a class party, Mr. Parker bought 8 pizzas for $64. How much did he spend for each pizza?

    $ ____ ÷ ____ = $ ____

4. Tony bought 60 pieces of candy. He wants to share the candy equally among his 6 friends. How many pieces of candy will he give each friend?

    ____ ÷ ____ = ____ candy pieces

5. A stack of 8 nature magazines costs $40. If each magazine has the same price, how much does each one cost?

    $ ____ ÷ ____ = $ ____

6. Carmen and José listened to 72 of their favorite songs. If there are 9 songs on each CD, how many CDs did they play?

    ____ ÷ ____ = ____ CDs

7. Toco Pet Shop Company received a shipment of 80 cartons of dog food. The same number of cartons will be sent to each of its 8 stores. How many cartons will each store receive?

    ____ ÷ ____ = ____ cartons

8. A group of 7 boys cut lawns during the weekend. Altogether, they made $56. Each boy will make the same amount. How much money did each boy earn?

    $ ____ ÷ ____ = $ ____

Name _____ Date _____

Dive into 45

Directions: Solve each problem.

A
4)8 2)6 5)10 7)14 8)72

B
3)12 4)4 8)16 6)18 7)49

C
1)10 2)8 4)20 9)27 5)35

D
9)0 5)5 5)15 10)20 6)36

E
10)50 5)20 2)2 6)48 9)72

F
3)18 6)60 9)18 8)24 7)63

G
10)60 7)28 4)12 9)36 9)45

H
4)16 3)15 6)24 8)56 1)4

I
2)20 5)25 3)21 2)16 10)90

Name _____   Date _____

Burst Past 45

Directions: Solve each problem.

A
10)80   2)4   6)30   3)24   10)30

B
3)6   6)6   6)54   8)40   7)42

C
5)40   4)28   9)54   3)30   1)3

D
9)63   4)32   7)56   6)42   5)45

E
2)10   4)24   7)35   8)80   3)3

F
8)48   7)7   3)27   4)36   10)40

G
9)81   3)9   7)21   2)18   9)9

H
4)40   5)30   6)12   8)32   2)14

I
8)64   5)50   8)8   2)12   1)8

KE-804077 © Key Education   53   Specific Skills: Division Facts Tips & Tricks

Classroom Division Games

The Nickel Trick Game

Materials Needed
- School play money: 50 pennies, 10 nickels, and 5 dimes for each player
- Paper and pencil

Getting Ready
Learning to divide by 5 is as easy as making change. Pair off students to play this game and provide them with the designated amounts of school play money.

How to Play
Introduce the ÷5 facts by telling students to think of 5 as 5¢, which can be shown with 5 pennies or 1 nickel. Review with students how to exchange 5 pennies for a nickel, 10 pennies for 2 nickels, and so on. Continue the explanation by discussing other division facts: "What is 15 ÷ 5?" What is 20 ÷ 5?" and "What is 25 ÷ 5?" Also, have students show the dividend with different coin combinations and then find the quotient (the number of nickels) for each given problem. For example, show the dividend as 2 dimes and 5 pennies. Direct students to figure out the total number of cents and then divide that amount by 5¢. Ask, "How many nickels is the answer?" (5 nickels) What is the division sentence? (25¢ ÷ 5¢ = 5 nickels)

Finally, it is time to play the game. Announce a division problem from the list below and have the players use their play money to solve the problem. Award a point to the first player who calls out the correct answer and shows the corresponding coins. Continue the game for a predetermined number of rounds.

Nickel Trick Game List
- 5¢ ÷ 5¢ = 1 nickel
- 10¢ ÷ 5¢ = 2 nickels
- 15¢ ÷ 5¢ = 3 nickels
- 20¢ ÷ 5¢ = 4 nickels
- 25¢ ÷ 5¢ = 5 nickels
- 30¢ ÷ 5¢ = 6 nickels
- 35¢ ÷ 5¢ = 7 nickels
- 40¢ ÷ 5¢ = 8 nickels
- 45¢ ÷ 5¢ = 9 nickels
- 50¢ ÷ 5¢ = 10 nickels

Division in a Bag

Materials Needed
- Paper bags, paints, and watercolor markers
- Index cards and pencils
- Story problems (page 57)

Getting Ready
Invite students to be creative by designing a grocery store bag, a variety store bag, a pet store bag, a barber shop bag, or another kind of bag. Using paint or markers, have students decorate their bags in their chosen themes. Then, challenge students to write a specified number of division word problems on index cards that relate to their theme bags or use some of the problems provided on page 57 in the bags.

How to Play
This game can become an activity center or classroom game. Have students individually exchange their division bags with classmates and then solve the math problems or work in teams to earn points.

Baseball Division Bingo

Materials Needed
- Game board (page 58)
- Bingo chips and watercolor markers
- Division flash cards

Getting Ready
Make one copy of the bingo board for each player.

How to Play
Give each student a copy of the bingo board. Have students choose 24 of the dividends listed at the bottom of their game boards and write them on the baseballs with markers. When students have completed filling in their game boards, distribute bingo chips to the players. Before starting the game, remove the 11's and 12's facts if they are included in the flash card set. Then, say aloud the division problems, one at a time, with the divisor and quotient but without the dividend (e.g., ___ ÷ 8 = 8). Whenever students have the missing dividend on their cards, they may cover that space with a bingo chip. Continue playing the game until one student has covered the predetermined spaces and calls out, "BASEBALL BINGO!"

Classroom Division Games

Division Bowling

Materials Needed
- Division flash cards
- White marker board
- Dry erase marker

How to Play
Select 10 students to represent bowling pins. Give each student (bowling pin) a few flash cards. Direct those students to stand in a pyramid shape. The rest of the class will be the bowlers and should form a line facing the "pins" (students). Use the white board/chalkboard to keep score.

The "head pin" (first student in the pyramid) will show Bowler A one flash card. Bowler A must correctly answer the division flash card in 10–12 seconds. If the answer is correct, the bowling pin (student) sits down and Bowler A receives one point and continues to answer problems. If the answer is incorrect, Bowler A moves to the back of the pyramid, the "head pin" moves to the back of the bowler line, and Player B answers the next problem. All of the bowling pins should be standing again for Bowler B. If a bowler "knocks down" 10 pins (answers 10 questions correctly), this is a strike! Play continues until all students have bowled.

Circle Division Game

Materials Needed
- Division flash cards (large cards)

How to Play
Ask students to sit in a circle (probably no more than 10–12 students). Then, choose a student to start the game. That student (Player A) stands behind the next player (Player B) who is sitting in the circle. The teacher holds up a flash card. The first student (Player A or B) to give the correct answer stands behind the next person (Player C) in the circle. (If Player B gave the correct answer first, Player A switches places with Player B by sitting down.) The game continues until at least one student makes it completely around the circle—thus, the Circle Division Game!

Division War!

Materials Needed
- 4 sets of index cards, numbered 0–10, for each group of 2 players

How to Play
Start the activity by shuffling the cards and dealing them facedown into two piles. The players keep their cards facedown in the piles. Count "1, 2, 3, Go!" On "Go," both players simultaneously turn two cards faceup in front of each other. The first player to use two or three of the four digits shown to make the dividend and the divisor of a math sentence and announce the correct quotient wins all of the cards played that round. (For example, if 2, 3, 5, and 8 are drawn, either the fact 32 ÷ 8 = 4 or 8 ÷ 2 = 4 can be called out. If neither player can make a division sentence, the players individually turn over one more card and try again to figure out a solution to win that round of play. Continue the game until one of the players no longer has a stack of cards.

Classroom Division Games

Put a Spin on It!

Materials Needed
- Blank spinner cards (see page 59)
- Dividend Bank (see page 59)
- Paper clip and pencil
- Scratch paper
- White marker board
- Dry erase marker

Getting Ready
Give each pair of students a copy of the blank spinner cards on page 59 along with a paper clip. Select eight dividends randomly (see page 59) and write them on the white board. Students can use the numbers to fill in one of their spinner cards.

How to Play
Begin the activity by choosing two dividends from the list on the board as the "targets" for players to use when stating division facts. Show students how to insert a pencil point through one end of the paper clip to make a spinner dial. On "Go!" have the players spin their paper clips. The players whose clips land on a target dividend must correctly state a related division fact to earn a team point. If more than one team can state a division fact for the target dividend, each pair of players must say a different fact to earn a point. For example, if 27 is one of the targets, a team may say 27 ÷ 3 = 9, 27 ÷ 9 = 3, 27 ÷ 27 = 1, or 27 ÷ 1 = 27. Continue the game for several rounds of play, selecting a different pair of target dividends each time, until one of the teams has earned the predetermined number of points to win. For the second game, have students fill in the other blank spinner card by using a new set of eight dividends.

Variation: Have students play against their partners. Allow two minutes to use each set of target dividends. The players take turns spinning the paper clip and recording a different division sentence on scratch paper.

Number Cube Division

Materials Needed
- 3 dice
- Paper and pencil

Getting Ready
Give each pair of students a set of three dice. Each player prepares a personal game board by writing the numbers 0–18 on a sheet of paper. Have the players circle those numbers to make them more prominent on their papers. Higher order thinking skills may entice students to use more than two mathematical operations as they play. This game lends itself to working with the order of operations.

How to Play
To begin the game, have Player A roll the dice and total the amount shown on them, using one or more operations (addition, subtraction, and/or multiplication). Player A states a division fact using that total as the dividend. The quotient is marked or crossed out on the paper and the generated math sentences should also be recorded near it.

For example, if Player A rolls 5, 6, and 3, the total would be 14 if addition is used. Player A could state 14 ÷ 2 = 7. The quotient of 7 is marked on the paper. If 7 has already been marked, that player must adjust the division fact to produce a different quotient, such as 14 ÷ 7 = 2. Player B continues the game in the same manner. As the game progresses, each player may have to be more creative to arrive at a quotient that remains on the paper. For example, if 5, 6, and 3 are rolled again, the player could think 6 x 3 = 18 and 18 x 5 = 90. Then, he might use the division fact 90 ÷ 10 to cross out the 9. The first player to cross out all numbers on his prepared paper wins the game.

Game directions on page 54

Division in a Bag

Jennifer hiked 54 miles on her camping trip. If she hiked 9 miles each day, how many days was she on her trip?	In the auditorium, there are 81 balcony seats. If there are 9 rows, how many balcony seats are in each row?
Paul has 30 tennis balls. A can holds 3 tennis balls. If he puts all of the balls in cans, how many cans will he fill?	Heath plays a drum in a band that has 81 members. There are 9 rows of players. How many players are in each row?
Blake purchased 45 designer pencils for his 9 friends. He gave each friend the same number of pencils. How many pencils did each friend receive?	A stack of 6 sports magazines costs $48 altogether. If their prices are the same, what is the cost of each magazine?
Cynthia has 25 handmade vases to sell at the art fair. If she arranges 5 vases in each row, how many rows will she have?	Carmen has 24 shoes in her closet. If each shoe is paired with its correct mate, how many pairs of shoes does she have?
Nikki bought 18 small books to give her friends as gifts. She gives each friend 3 books. How many friends were given books?	Sonja is making 28 holiday candles for her friends. She will give each friend 4 candles. How many friends will receive candles?
Every morning, Mateo helps his brother deliver the newspaper in his town. He delivers a total of 63 newspapers in one week. How many customers does his brother have?	White's Pet Store's warehouse received 60 boxes of bird seeds. The same number of boxes must be sent to each of 6 stores. How many boxes of bird seed will each store get?
Brenda is making 100 cookies for a birthday party. She can bake 10 cookies on her baking sheet. How many times does she fill the baking sheet?	Barry and Gail have 9 hamsters. They plan to sell each one for the same price to earn $90. How much will each hamster cost?

KE-804077 © Key Education — Specific Skills: Division Facts Tips & Tricks

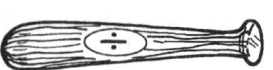
Baseball Division
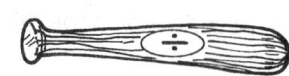

Game directions on page 54

B	I	N	G	O
		FREE SPACE		

Directions: Choose 24 numbers. Using a marker, write each number on a baseball.

Dividend Bank

0	5	10	18	27	36	49	63	81
1	6	12	20	28	40	50	64	90
2	7	14	21	30	42	54	70	
3	8	15	24	32	45	56	72	
4	9	16	25	35	48	60	80	

Specific Skills: Division Facts Tips & Tricks

Game directions on page 56

Put a Spin on It!

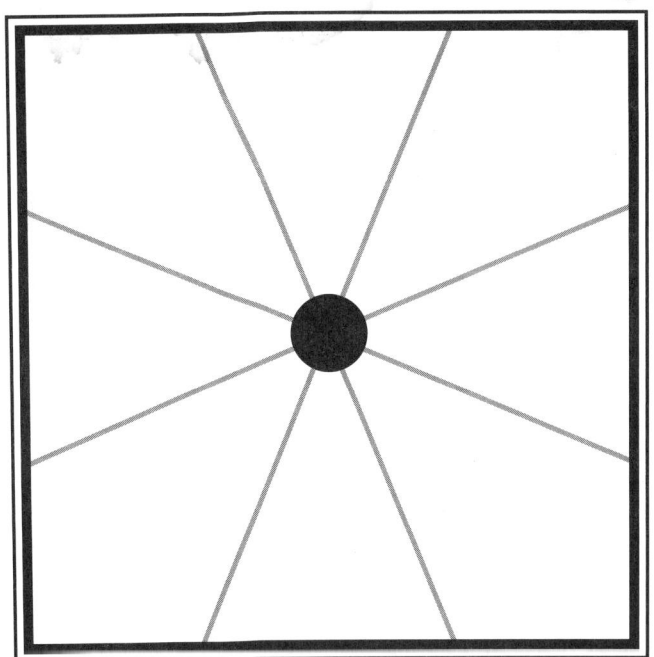
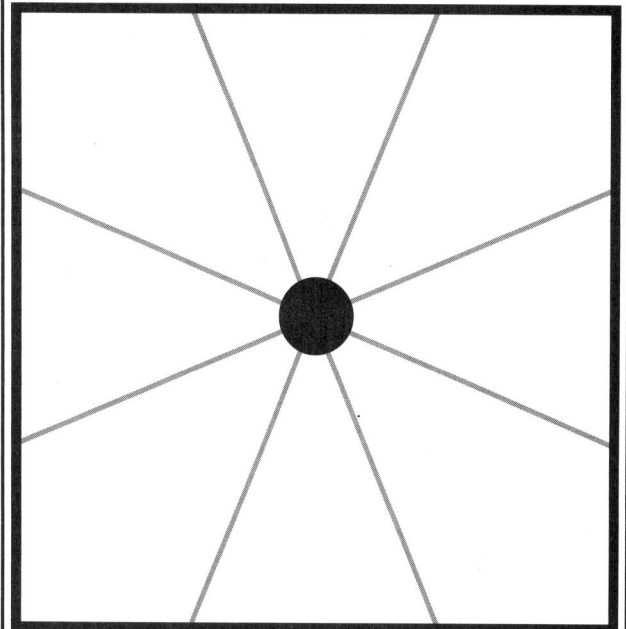

Put a Spin on It!

✂ -

Dividend Bank

☐ 0	☐ 12	☐ 30	☐ 56
☐ 1	☐ 14	☐ 32	☐ 60
☐ 2	☐ 15	☐ 35	☐ 63
☐ 3	☐ 16	☐ 36	☐ 64
☐ 4	☐ 18	☐ 40	☐ 70
☐ 5	☐ 20	☐ 42	☐ 72
☐ 6	☐ 21	☐ 45	☐ 80
☐ 7	☐ 24	☐ 48	☐ 81
☐ 8	☐ 25	☐ 49	☐ 90
☐ 9	☐ 27	☐ 50	
☐ 10	☐ 28	☐ 54	

Tic-Tac-Division

To the Teacher: Tic-Tac-Division is another great way to review division facts. Make a copy of the template below. Write some division problems both horizontally as well as vertically on the grid. Then, duplicate the game board for students to use.

To Play: Have students play Tic-Tac-Division in pairs or in small groups. Students can play the game the traditional way or you may consider having them use the following rules:

- ✗ Students can flip a coin to see who will start the game and choose the symbol *X* or *O*.
- ○ When a problem is solved correctly, the player marks the square with the chosen letter.
- ✗ Players check each other's answers.
- ○ In the event that a division fact is answered incorrectly, the player may not mark the square. The opposing player then takes a turn and may mark that space or choose another problem to solve.
- ✗ A player must correctly complete three problems in a row either horizontally, vertically, or diagonally to win.
- ○ To finish the game, the player must solve the problem correctly on the winning game square by writing down the answer without changing it.

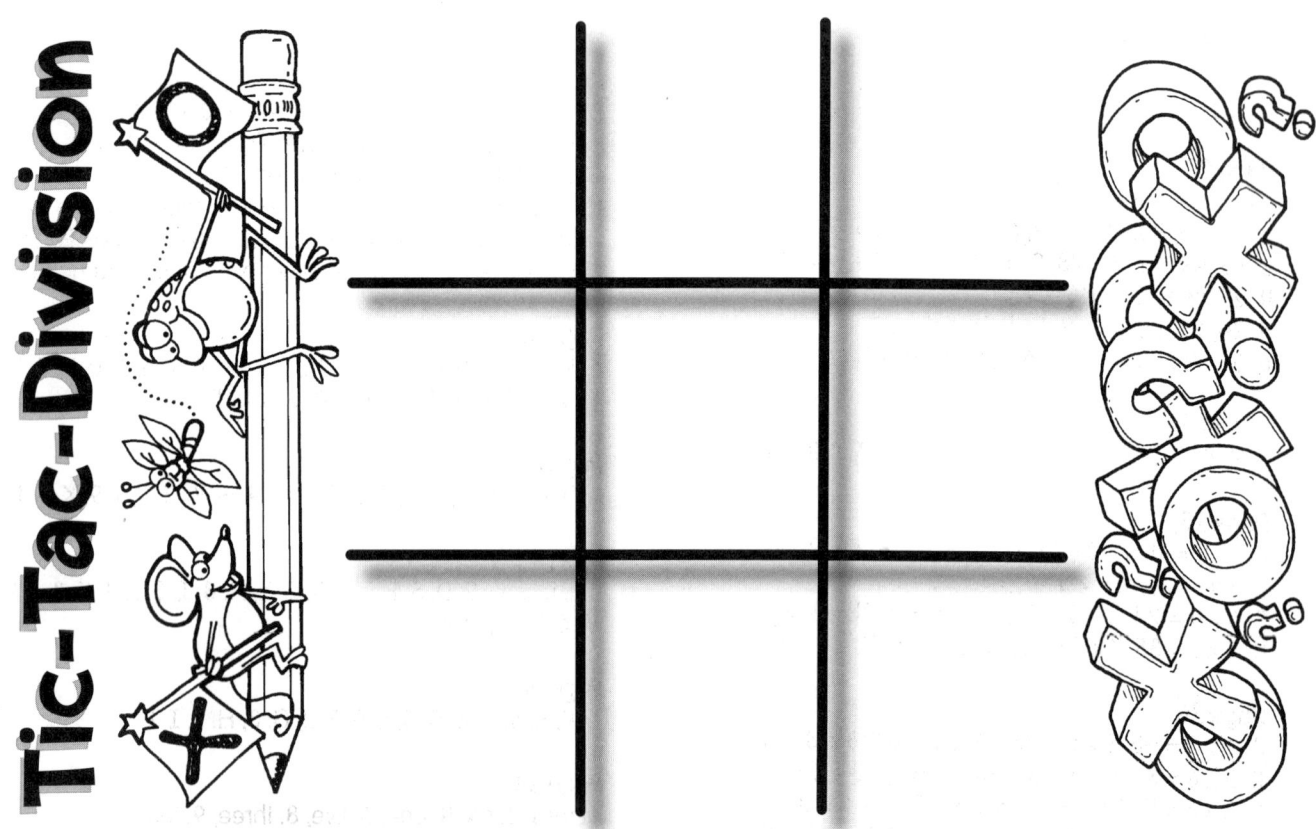

Answer Key

Page 5
1. B, 2. C, 3. C, 4. A, 5. B, 6. C

Page 6
1. B, 2. C, 3. B, 4. C, 5. A, 6. B

Page 17
1. 3, 12 ÷ 4 = 3; 2. 4, 12 ÷ 3 = 4;
3. 7, 14 ÷ 2 = 7; 4. 5, 15 ÷ 3 = 5

Page 18
1. 3, 18 ÷ 6 = 3; 2. 3, 21 ÷ 7 = 3;
3. 5, 20 ÷ 4 = 5; 4. 4, 32 ÷ 8 = 4

Page 19
1. 7, 28 ÷ 4 = 7; 2. 4, 16 ÷ 4 = 4;
3. 5, 10 ÷ 2 = 5; 4. 4, 24 ÷ 6 = 4

Page 20
1. 5, 30 ÷ 6 = 5; 2. 5, 20 ÷ 4 = 5; 3. 6, 18 ÷ 3 = 6;
4. 4, 28 ÷ 7 = 4; 5. 3, 15 ÷ 5 = 3; 6. 7, 35 ÷ 5 = 7

Page 21
1. A, 7; 2. C, 4; 3. E, 6; 4. H, 2; 5. F, 1; 6. B, 4; 7. D, 4; 8. G, 2

Page 22
A. 3, 9 ÷ 3 = 3; B. 3, 18 ÷ 6 = 3;
C. 9, 36 ÷ 4 = 9; D. 8, 24 ÷ 3 = 8

Page 23
1. 10; 50 ÷ 5 = 10
2. 42, 36, 30, 24, 18, 12, 6, 0; 8; 48 ÷ 6 = 8
3. 56, 49, 42, 35, 28, 21, 14, 7, 0; 9; 63 ÷ 7 = 9
4. 27, 18, 9, 0; 4; 36 ÷ 9 = 4
5. 24, 20, 16, 12, 8, 4, 0; 7; 28 ÷ 4 = 7
6. 56, 48, 40, 32, 24, 16, 8, 0; 8; 64 ÷ 8 = 8

Page 24
A. 8, B. 6, C. 2, D. 6, E. 9, F. 6, G. 6, H. 7, I. 8, J. 8, K. 4, L. 8

Page 25
1. 2 x 5 = 10, 5 x 2 = 10, 10 ÷ 2 = 5, 10 ÷ 5 = 2
2. 8 x 1 = 8, 1 x 8 = 8, 8 ÷ 1 = 8; 8 ÷ 8 = 1
3. 6 x 4 = 24, 4 x 6 = 24, 24 ÷ 6 = 4, 24 ÷ 4 = 6
4. 5 x 7 = 35, 7 x 5 = 35, 35 ÷ 5 = 7, 35 ÷ 7 = 5

Page 26
A. 6 x 8 = 48, 8 x 6 = 48, 48 ÷ 6 = 8, 48 ÷ 8 = 6
B. 5 x 6 = 30, 6 x 5 = 30, 30 ÷ 5 = 6, 30 ÷ 6 = 5
C. 3 x 5 = 15, 5 x 3 = 15, 15 ÷ 3 = 5, 15 ÷ 5 = 3
D. 5 x 8 = 40, 8 x 5 = 40, 40 ÷ 5 = 8, 40 ÷ 8 = 5
E. 6 x 7 = 42, 7 x 6 = 42, 42 ÷ 6 = 7, 42 ÷ 7 = 6
F. 7 x 4 = 28, 4 x 7 = 28, 28 ÷ 7 = 4, 28 ÷ 4 = 7
G. 8 x 3 = 24, 3 x 8 = 24, 24 ÷ 8 = 3, 24 ÷ 3 = 8
H. 9 x 2 = 18, 2 x 9 = 18, 18 ÷ 9 = 2, 18 ÷ 2 = 9
I. 3 x 9 = 27, 9 x 3 = 27, 27 ÷ 3 = 9, 27 ÷ 9 = 3
J. 4 x 9 = 36, 9 x 4 = 36, 36 ÷ 4 = 9, 36 ÷ 9 = 4

Page 27

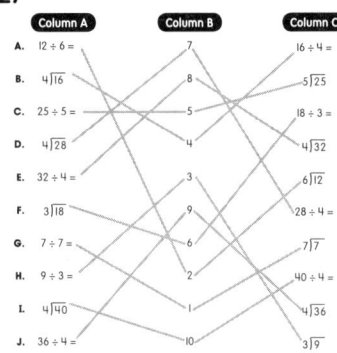

Page 28
Circled facts: B. 24 ÷ 6 = 4; C. 63 ÷ 7 = 9; D. 15 ÷ 3 = 5;
G. 16 ÷ 4 = 4; I. 81 ÷ 9 = 9; K. 56 ÷ 8 = 7; L. 72 ÷ 9 = 8
Corrected facts: 6 ÷ 6 = 1, 27 ÷ 3 = 9, 30 ÷ 5 = 6,
45 ÷ 5 = 9, 18 ÷ 6 = 3
M. 7, N. 7, O. 2, P. 4, Q. 7, R. 5, S. 5, T. 8, U. 6, V. 7, W. 8

Page 29
A. 2, 4, 8 ÷ 2 = 4; B. 4, 4, 16 ÷ 4 = 4; C. 3, 8, 24 ÷ 3 = 8;
D. 4, 8, 32 ÷ 4 = 8; E. 2, 10, 20 ÷ 2 = 10, F. 1, 14, 14 ÷ 1 = 14
Bottom of page: A. 5, B. 4, C. 9, D. 2, E. 6, F. 2, G. 4,
H. 1, I. 2, J. 4, K. no answer, L. odd, M. even
For the basic facts, the "even ÷ odd = even" and the "odd ÷ odd = odd" patterns always work. The answers for the "even ÷ even" facts will be either even or odd numbers.

Page 30
A. 4, B. 0, C. 9, D. 1, E. 0, F. 12, G. 0, H. 6, I. 5, J. 10, K. 3, L. 0,
M. 7, N. 8, O. 0, P. 9, Q. 0, R. 0, S. 0, T. 0

Page 31
A. 4, B. 5, C. 9, D. 7, E. 6, F. 8, G. 4, H. 10, I. 7, J. 9, K. 6, L. 8

Page 32
A. 1, B. 2, C. 2, D. 5, E. 7, F. 5, G. 7, H. 8, I. 3, J. 9, K. 3, L. 4,
M. 6, N. 10, O. 10

Page 33
A. 4, B. 5, C. 6, D. 8, E. 9, F. 2, G. 3, H. 7, I. 1

Page 34
Across: 3. six, 4. one, 6. five, 8. three, 9. ten
Down: 1. two, 2. nine, 5. eight, 6. four, 7. seven

Answer Key

Page 35
A. 4, B. 3, C. 8, D. 5, E. 6, F. 9, G. 7, H. 5, I. 8, J. 3, K. 9, L. 4, M. 7, N. 6

Page 36
A. 5, B. 4, C. 7, D. 5, E. 8, F. 4, G. 7, H. 6, I. 9, J. 2, K. 1, L. 9, M. 1, N. 3, O. 10

Page 37
Multiples of 6 shaded: 6, 12, 18, 24, 30, 36, 42, 48, 54, 60
A. 6, B. 2, C. 3, D. 8, E. 7, F. 9, G. 4, H. 5
Multiples of 7 shaded: 7, 14, 21, 28, 35, 42, 49, 56, 63, 70
I. 6, J. 3, K. 2, L. 4, M. 8, N. 5, O. 9, P. 7

Page 38
Multiples of 8 shaded: 8, 16, 24, 32, 40, 48, 56, 64, 72, 80
A. 8, B. 2, C. 6, D. 7, E. 5, F. 3, G. 9, H. 4
Multiples of 9 shaded: 9, 18, 27, 36, 45, 54, 63, 72, 81, 90
I. 2, J. 9, K. 8, L. 4, M. 7, N. 5, O. 3, P. 6

Page 39
A. 2, 5, 8, 3, 7, 6, 10, 10
B. 5, 9, 5, 10, 6, 7, 9, 3
C. 0, 2, 4, 2, 4, 9, 7, 1
D. 6, 2, 6, 1, 0, 3, 9, 8
E. 4, 8, 3, 4, 5, 7, 6, 8

Page 40
1. 12 ÷ 6 = 2; 2. 32 ÷ 8 = 4; 3. 27 ÷ 3 = 9; 4. 72 ÷ 9 = 8; 5. 18 ÷ 3 = 6; 6. 54 ÷ 6 = 9; 7. 42 ÷ 7 = 6; 8. 80 ÷ 8 = 10

Page 41
A. 2, B. 6, C. 3, D. 8, E. 4, F. 10, G. 7, H. 5, I. 1, J. 0, K. 9, L. 10, M. 10, N. 10, O. 60, P. 10, Q. 10, R. 10, S. 80, T. 10, U. 4, V. 10, W. 10, 80 ÷ 8 = 10

Page 42
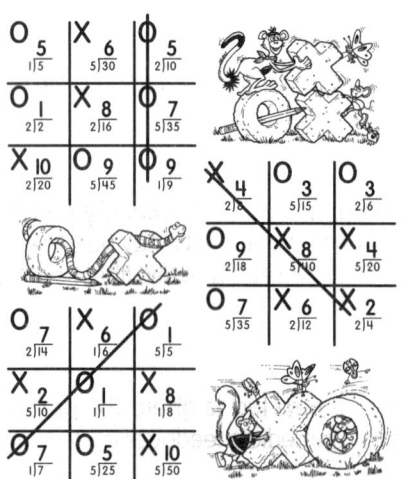

Page 43

Page 44
A. 8, 3, 4, 0, 7, 2
B. 6, 0, 10, 5, 9, 1

Page 45
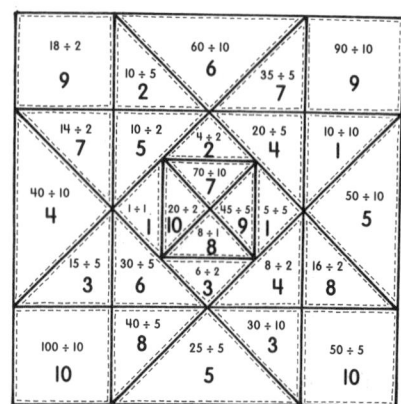

Page 46
Start to Finish: 6, 5, 3, 7, 7, 9, 3, 8, 6, 2, 8, 9, 5, 2, 4, 4

Page 47
A. 21, B. 24, C. 12, D. 40, E. 27, F. 16, G. 30, H. 32, I. 18, J. 36

Page 48
A. 4, 48, 7, 60, 5, 18, 6, 1, 12, 9
B. 7, 18, 4, 10, 45, 9, 8, 81, 3, 54

Page 49
A. 7, B. 10, C. 5, D. 1, E. 2, F. 6, G. 9, H. 3, I. 8, J. 8, K. 5, L. 6, M. 4, N. 9, O. 3, P. 1, Q. 7, R. 4, S. 10, T. 2

Answer Key

Page 50

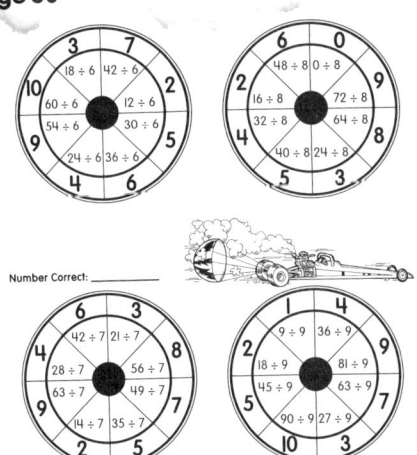

Page 51
1. 21 ÷ 7 = 3; 2. $81 ÷ 9 = $9; 3. $64 ÷ 8 = $8;
4. 60 ÷ 6 = 10; 5. $40 ÷ 8 = $5; 6. 72 ÷ 9 = 8;
7. 80 ÷ 8 = 10; 8. $56 ÷ 7 = $8

Page 52
A. 2, 3, 2, 2, 9; B. 4, 1, 2, 3, 7; C. 10, 4, 5, 3, 7; D. 0, 1, 3, 2, 6;
E. 5, 4, 1, 8, 8; F. 6, 10, 2, 3, 9; G. 6, 4, 3, 4, 5; H. 4, 5, 4, 7, 4;
I. 10, 5, 7, 8, 9

Page 53
A. 8, 2, 5, 8, 3; B. 2, 1, 9, 5, 6; C. 8, 7, 6, 10, 3; D. 7, 8, 8, 7, 9;
E. 5, 6, 5, 10, 1; F. 6, 1, 9, 9, 4; G. 9, 3, 3, 9, 1; H. 10, 6, 2, 4, 7;
I. 8, 10, 1, 6, 8

Web Sites

http:www.aplusmath.com/flashcards/division.html
When additional practice is needed, check out the division flash cards offered on this Web site. The player is given good immediate feedback for correct and incorrect answers. After answering 10 facts, a summary screen shows the results: number correct, number of attempts, score, and message. In the game room, students may be interested in playing Division Matho.

http://www.amblesideprimary.com/ambleweb/mentalmaths/dividermachine.html
Players choose the level of difficulty and then answer the division fact when playing Ambleweb Division Machine. Feedback is provided for correct and incorrect answers. The game works well for students who need extra time for solving math facts.

http:www.learn-with-math-games.com
Have a little fun with math or have a lot of fun. Either way, use teacher-approved math games to motivate and teach math concepts, including division. The Web site offers an extensive list of links for math games that include Ambleweb Division Machine and SumSense.

http://resources.oswego.org/games/SumSense/sumdiv.html
Drag and drop numeral cards to create division sentences that make sense. Players only have two minutes to solve eight problems to beat the clock. A great game for challenging those students who know the facts!

http:www.sheppardsoftware.com/math.htm
This Web site offers online learning games for the basic math operations. Students play a game individually and may compete against other scores that have been posted in the network. To practice division facts, students may be interested in playing Fruit Shoot, Pop-up Math (called Croc Doc, focusing on specific fact families), and Matching Mania (the more facts a player answers correctly and quickly, the more points the player earns).

http:www.mathplayground.com/division01.html
When playing this game, there are several chances to answer each problem before a new fact is given. The problems appear in random order up to the ÷ 12 family and are shown in the long division format. Immediate feedback is given when each answer is checked.

Correlations to NCTM Standards

Division Facts Tips & Tricks supports the NCTM *Principles and Standards for School Mathematics* (2000).

The activities in this book support the following Number and Operations Standard Expectations for Grades 3–5:

1. **Students recognize equivalent ways to represent the same number and generate equivalent numbers by composing or decomposing.**
   In this book, students learn that a number can be represented by different division problems.

2. **Students understand the meanings of multiplication and division.**
   The focus of this book is learning division facts, so the activities in it support students in understanding the meaning of division.

3. **Students understand the effects of multiplying and dividing whole numbers.**
   All activities in this book support this standard.

4. **Students identify relationships between operations and use these relationships, such as division being the inverse of multiplication, to solve problems.**
   Certain activities in this book present division as repeated subtraction or as the inverse of multiplication, so they support this standard.

5. **Students understand and use properties of operations.**
   In this book, students learn about the zero and identity properties of division.

6. **Students become fluent in basic number combinations for multiplication and division and use these to compute related problems mentally.**
   The focus of this book is developing fluency in basic division facts.

7. **Students become fluent in adding, subtracting, multiplying, and dividing whole numbers.**
   This book supports students becoming fluent in basic division.

8. **Students select appropriate computation methods and tools from among mental math, estimation, and paper and pencil according to the context and then use the selected methods and tools.**
   This book presents a wide variety of methods and tools for learning division facts and enables students to learn when to choose particular methods.

The activities in this book support the following Algebra Standard Expectations for Grades 3–5:

1. **Students describe, extend, and generalize about geometric and numeric patterns.**
   Skip counting and array activities in this book, which form a basis for division, support this standard.

2. **Students model problems with objects and use graphs, tables, and equations to draw conclusions.**
   Students use pictures and arrays to model division in several activities in this book.